TX 946 .G5 1980

Applied cook-freezing

DATE DUE

APPLIED COOK-FREEZING

APPLIED
COOK–FREEZING

Compiled by

PETER GLANFIELD

Head, Residential Services Department,
University of Keele, Staffordshire, UK

APPLIED SCIENCE PUBLISHERS LTD
LONDON

APPLIED SCIENCE PUBLISHERS LTD
RIPPLE ROAD, BARKING, ESSEX, ENGLAND

British Library Cataloguing in Publication Data

Glanfield, Peter
  Applied cook–freezing.
  1. Cookery for institutions, etc.—Case studies
  2. Food, Frozen—Case studies
  3. University of Keele
  I. Title
  642'.58   TX946

  ISBN 0-85334-888-X

WITH 2 TABLES AND 32 ILLUSTRATIONS

© APPLIED SCIENCE PUBLISHERS LTD 1980

Printed in Great Britain by Galliard (Printers) Ltd, Great Yarmouth

Preface

The problem of providing residential accommodation together with a satisfactory catering service to large numbers of students at a price that they can afford and are prepared to accept without creating deficits is one that has increasingly perplexed university administrators in recent years. Whilst the University of Keele undoubtedly has certain advantages both in its location and in its overall administrative structure the problem of providing a self-balancing residential service to students is more acute than at most other universities simply because Keele is a campus university with the vast majority of students and staff resident on site. Some of the problems at Keele are described later together with the ways in which some have been overcome. Clearly this book is mainly concerned with the practical management of catering by the application of imaginative systems. The introduction and proving of these systems have been a long and often agonised process and it would obviously be impossible within the space available to closely chronicle the whole of the process. Instead it has been decided to simply describe some of the main events in an attempt to provide signposts for others who may be contemplating taking a similar course of action. There has also been little attempt to provide scientific or technical details on the cook–freezing process as other publications already amply cover these areas.

A significant number of visitors with catering problems visit Keele each year in the hope that they will see something to help them out of their difficulties. Unfortunately there are no instant panaceas in catering and although we make as much information as possible available to these visitors it is up to each and every one of them to adapt and modify our experience to fit their own requirements. In essence this book is the information that we give to these visitors.

In passing, it is perhaps of some interest to mention that the application of cook–freezing is not at all new to Keele. The earliest record of this technique that we have can be found in a letter written by the 11-year-old John Strange Spencer Churchill to his elder brother Winston Churchill during a stay at Keele during December 1891. He wrote 'It is beautiful here, the sun is out etc., and freezing hard ... at breakfast this morning the gardener found a poor frozen fish which was on the ice . . .'. In those days the fish would have been reheated or 'convected' in the kitchen by the family's French cook, who, the records state, was paid an annual salary of £240.

Although Keele no longer has a catering or residential problem either in terms of financial viability or customer acceptance, it must be admitted that the catering systems that have been introduced are just one part of the answers that have been found. There are, of course, many other factors and perhaps the principal of these is the development of the vacation-letting trade. The academic year consists of three ten-week terms and these may be regarded as the peaks, leaving the remaining twenty-two weeks in each year as the troughs of activity. In the past there were times when large resources of labour and facilities were significantly under-utilised. This trough obviously had to be filled and as a result steps were taken to attract conferences and meetings to the university with a view to creating additional revenue by utilising under-used resources. This policy has, over the years, been so successful that the University of Keele has now become an extremely active conference centre hosting nearly two-hundred groups each year and selling some 80 000 bednights during twenty weeks of vacation each year. Income from this source is now considerable and although its principal objective is to ensure that student residency fees are kept as low as economically possible it also allows the university to subsidise certain uneconomic but socially necessary services and to provide capital sums for the maintenance and improvement of residential facilities. By this intelligent use of spare capacity and by application of modern catering and management techniques Keele University has been able to create a situation in which residency fees are amongst the lowest in Great Britain.

Contents

Preface v
Acknowledgements ix
Introduction xi

Part I—The Feasibility Study

Introductory Remarks 3
Visits 3
The Old and the New—a Comparison 5
Conclusion and Go-ahead 7
The Tender 7
Installation Programme 11
Initial Assessment for the Period to 31 December 1973 . . 12
Refectory Services Final Work Study Report 12
Appendices 29

Part II—Recipes

Introductory Remarks 45
Choice of Ingredients 45
Degree of Cooking 47
Portioning 47
Standard Recipes 48
Keele Recipes 49
Meat 51
Chicken, Duck and Game 83
Fish 97
Farinaceous and Vegetarian 117
Sweets 135

Part III—Catering Costing System

Introductory Remarks 159
Objectives of the System 159
The Computer System 161
Input Documents 161
Computer Reports 174

Index 197

Acknowledgements

Any management system that can be successfully applied to a large organisation must of necessity result from the efforts of a large number of individuals. Keele systems have been evolved over a number of years and although certain individuals have made particularly significant contributions the real credit must be attributed to team work. Whilst the enthusiasm and effort of the catering and domestic staff must be acknowledged there is no doubt that the enlightened encouragement given to line management by the senior officers of the university has also been a major contributory factor in any success that has been achieved. It is unfortunately not possible to do more than list below some of the university staff whose ideas and work have contributed to the production of this book.

The Cook–Freeze Project
Stephen Ware   Now at the University of Kent at Canterbury.
Albert J. Murden   Now retired.
Kenneth Riley   Organisation & Methods and Work Study Officer
                      —University of Keele.

Recipe Development
Malcolm Griffiths   Production Manager—University of Keele.

Stock and Financial Control Systems
Noel Walley   Data Processing Officer—University of Keele.
David Thompson   Now at the Milk Marketing Board.
Derek Tomkinson   Residential Services Department—University of Keele.

Introduction

The University of Keele is situated in North Staffordshire some $3\frac{1}{2}$ miles from Newcastle-Under-Lyme. The university estate of 650 acres, with extensive woods, lakes and parkland, was formerly owned by the Sneyd family.

The University College of North Staffordshire was founded in 1949, becoming the University of Keele in 1962. There are both three- and four-year courses for undergraduates at Keele. In the four-year course, the first year is known as the Foundation Year, which has been described as 'the most original innovation in British university education this century'. In the foundation year, students follow a broad course covering the development of Western civilization. The core of the course is a series of lectures to which all University departments contribute, and another feature is that students attend tutorial classes in nine different subjects.

Having deliberately chosen to expand slowly, Keele remains one of the smallest universities in Britain with a current planning objective of 3200 full time registered students in 1981/82. The residency task is, however, unique in that all but approximately 250 of these students, together with a high proportion of the staff, reside on the campus. Of the students who live on campus nearly three quarters are resident in University Grant Committee financed traditional Halls of Residence. These students dine in the University refectories and are required to pay for ten meals a week for thirty weeks each year. The remainder of the students live in loan-financed self-catering flats which they occupy on a 48-week tenancy agreement.

Three of the four Halls of Residence have modern kitchens and dining rooms equipped to produce a total of 50 000 meals a week. The catering service, which is the responsibility of the Residential Services Department, is managed centrally and each of the three refectories operates to the same menu and conforms to the one-management criteria.

Before 1968 every student resident on the campus received accommodation and twenty meals each week. However, since this date the pattern of student demand for catering facilities has changed significantly and has resulted in a gradual decline in the number of meals served each week. By the end of 1970 a situation had been reached where only ten meals per week were being served to each resident student as part of the residence fee. By January 1971, facilities that were capable of serving 50 000 meals a week were reduced to a weekly output of only 15 500 meals even though during this period the catering operation was the focus of almost continuous experiment, investigation and revision in an attempt to ensure that the task was being carried out as effectively as possible. It soon became evident that the total income from students during term, and from conferences during vacations, was not sufficient to ensure adequate recovery of operating costs. With the emphasis on economy, reductions in service and the necessity to maintain as constant the level of student fees in the face of rising costs, there seemed little more that could be done to prevent the inevitability of reduction in standards and the efficiency of the operation.

In August 1969, management consultants were asked to advise on 'Potential economies in catering'. In July 1970, consultants were asked to advise on 'Future development of the university catering service', and for five months the university's own staff experimented with 'all-in', 'pay-as-you-go', 'contributory meal payment scheme' and other methods of reducing catering costs, whilst at the same time endeavouring to meet the new pattern of student demand. In December 1970, an unexpected national pay award added $18\frac{1}{2}\%$ to the cost of catering wages.

The management and catering consultants' reports referred to centralisation of food preparation as a means of reducing operating costs and improving general efficiency. This recommendation was examined and it soon became apparent that the central preparation and transportation of conventionally cooked food is difficult to achieve without loss of nutritional value, quality and variety. By the end of 1970, developments in the techniques of freezing and reconstituting of food had been researched and developed that seemed to overcome many of the difficulties associated with centralised food preparation and distribution, and investigations were begun to apply these to Keele University's catering task.

PART I

The Feasibility Study

INTRODUCTORY REMARKS

This study, which began during the first half of 1971, was carried out by a team of three: the then Residential Services Manager, S. C. Ware; the then Catering Officer, A. J. Murden; and the Organisation & Methods and Work Study Officer, K. J. Riley.

The study considered the conventional methods of catering and compared this with a scheme for the processing of pre-cooked frozen foods. Throughout the investigation great emphasis was placed on ensuring that standards of nutrition, quality, choice and menu variety were maintained. Terms of reference were agreed and were as follows:

1. To estimate the capital expenditure, annual running costs and benefits of changing the conventional catering system to a cook–freeze meals system.
2. To evaluate the practical and financial application of introducing a cook–freeze system.
3. To consider two schemes:
   (a) cook–freezing all possible dishes for lunch and dinner;
   (b) cook–freezing two courses, principal dish only for lunch and dinner.

VISITS

In order to gather information at the beginning of the study a number of visits were made to establishments already using the cook–freeze system. Each proved of value and provided evidence that a system of centralised cook–freeze food production was practical, economical and relevant to the Keele situation.

Information and advice made available during all the visits was a major factor in the decision made subsequently to proceed without the installation of a pilot scheme. Visits were made as follows:

Leeds University—Department of Food and Leather Technology

A cook–freeze pilot scheme had been operated in a Leeds hospital for a period of four years and showed that economies could be achieved without the loss of nutritional value, quality or variety.

Darenth Park Hospital in Kent

Catering for 1800 resident patients and another 900 patients in a nearby hospital. Little economy in the cost of raw materials was achieved, but a considerable saving was shown as a result of reduced labour force. It was estimated that at the rate of saving on labour costs alone, capital expenditure and additional running costs would have been recovered in $3\frac{1}{2}$ years from implementation.

Liverpool School Meals Organisation

A Department of Education and Science joint scheme with the Liverpool City Council instituted in three stages to cover all schools catering in the city over a period of 10 years. At the time of the visit the first stage had been in progress for three years and would eventually provide 25 000 lunches daily to selected primary, junior and secondary schools. Frozen products were bought in from commercial suppliers and shortly after our visit the second stage of development was due to be commenced with the appointment of a Production Manager and the opening of a central production kitchen. The third stage would double the first stage of 25 000 meals once all schools had been incorporated into the cook–freeze programme.

It was found that the school meal's image had improved considerably and the children's acceptance of cook–freeze lunches was evident in all the schools visited. Other benefits were improved working conditions and hours, which in turn attracted reliable staff more readily. Economy of staff employed where the cook–freeze system operated showed that as much as 2p per school meal could be saved.

Chester—Midlands Electricity Board Research Centre

Research into equipment design, nutrition values, etc., had been carried out and a trial plant installed at the Centre's works canteen. Savings in labour costs had been achieved, as had general acceptance without difficulty by canteen customers.

Ardeer, Scotland—ICI Factory

A pilot scheme started in 1968; the system showed a saving of 25 % in labour costs could be achieved. The programme was intended as a means of establishing equipment necessary and establishment level required to meet specific demand. A visit was made at a later date following acceptance of the Keele scheme to study kitchen production activities, costings, etc.

THE OLD AND THE NEW—A COMPARISON

Conventional Catering

Prior to the implementation of the cook–freeze system at Keele, food was prepared conventionally in three kitchens each associated with a Hall refectory. This method involved kitchen staff in peak activity before each meal and slack periods over other parts of the day. To facilitate the function of the refectory kitchens throughout the whole day a two-shift system of working was in operation.

An analysis of term and vacation food consumption showed the number of meals taken to be:

	Breakfast	Lunch	Dinner	Conference specials	Total
Term: (30 weeks)	(à la carte)	215 100	190 200	—	405 300
Vacations:					
Christmas	2 375	5 450	2 670	700	11 195
Easter	9 630	16 325	10 049	2 612	38 616
Summer	16 865	33 435	18 426	1 574	70 300
Total:	28 870	270 310	221 345	4 886	525 411

Whilst menus in the refectories were similar each day, according to a five-week menu cycle, the preparation and cooking was undertaken individually.

Menu planning was restricted to dishes which could be prepared and completed within the time available. It was often difficult to successfully include in the menu more than one dish requiring last-minute attention and decoration.

The provision of kitchen equipment took account of the production required to meet the comparatively short period of maximum demand prior to each meal. Consequently, equipment available was under-utilised for a large proportion of the period of daily service.

The Choice of Schemes

The two schemes considered were as follows:

Scheme 1

Pre-cooking and freezing of all possible dishes for lunch and dinner throughout the year. This system necessitates high capital expenditure on equipment and cold-storage space, whilst achieving the best economy and ensuring adequate variety.

A large cold-storage area enables greater flexibility of:

(a)   Variety of dishes produced.
(b)   Quantity of packs produced in advance of requirements.
(c)   The mass production of individual dishes requiring special attention at the time of production.
(d)   The opportunity to take advantage of seasonal bulk purchases.

Scheme 2

Pre-cooking and freezing of two dishes, principal dish only; the remaining requirements for lunch and dinner being prepared and cooked in the conventional manner.

Although capital outlay and equipment and cold storage are proportionately reduced, there are limitations which reduce effectiveness.

Difficulties arise in maintaining an even level of production throughout the year, due to lack of cold-storage space, which would otherwise act as a buffer between a maintained level of production throughout the year, and variable demand.

The feasibility study also included details of:

(a)   Calculations to show total production throughput necessary to maintain existing choice and variety.
(b)   Graphs to show distribution of actual annual consumption and production levels necessary to be achieved for either scheme in order to maintain adequate supply throughout the year.
(c)   Estimated cost comparison showing annual savings for both schemes.

CONCLUSION AND GO-AHEAD

The study concluded by recommending the choice of Scheme 1 on the grounds that it would:

(a)   Improve quality and variety of menus.
(b)   Simplify production control.
(c)   Provide greater flexibility and increased level of production.
(d)   Reduce the number of skilled staff employed in refectory kitchen.

It was estimated that:

(a)   Capital cost of implementation would be £30 000 and that an annual saving in the cost of labour of £13 000 could be achieved.
(b)   After allowing for additional running costs the net annual saving would be £6700.

At a meeting of the University's Catering Consultative Group on Saturday, 26 June 1971, the feasibility study was presented and the Group agreed to recommend for consideration that Scheme 1 be implemented.

Following a visit to the Midlands Electricity Board Research Centre by members of the university staff and students, approval was given by the Halls of Residence Committee at its meeting on 9 December 1971 for the implementation of Scheme 1 subject to approval by the Medical Officer of Health.

On 31 December 1971, the Medical Officer of Health wrote stating:

'... I am writing to confirm that insofar as this development is concerned no possible objections could be made to carrying out of the proposed revised catering scheme. In fact I consider that a scheme such as that proposed would include advantages in the field of prevention of food-borne infections.'

THE TENDER

At the end of January 1972, the full tender specification and conditions of tender were compiled and offered to three reputable companies who were experienced in the installation of cook–freeze equipment.

The tender specification included the detailed specification reproduced in Table 1. The form of detailed specification invited some criticism at a later stage, particularly when tenders were being compared prior to awarding the contract. However, much thought was given to the best method of achieving

TABLE 1
Detailed Specification

Each section to be priced separately

Section I
Preparation of site:
Remove existing equipment as necessary including:

(a) four electric ovens,
(b) one open gas cooking range,
(c) one Hobart mixer,
(d) one free-standing fridge,
(e) one ice-making machine,
(f) one sink unit,
(g) one wash-hand basin,
(h) all stainless steel tabling,
(i) all other items considered necessary to complete preparation of site to be stated in the tender.

Section II
Alterations to building structure and services:
Give details of alterations and additional works considered necessary to enable total installation of proposed scheme.

Section III
Equipment for blast-freezing, cold storage and reheating:
(a) Blast-freezer and central cold store:

To be sited in existing Lindsay Refectory kitchen area.
Give details of equipment considered capable of complying with the demand and menu cycle.
It is envisaged that the kitchen production unit will operate an eight-hour day, five days per week.
State clearly how capacity of scheme recommended has been determined.

(b) Outlet storage and reheating:

The three refectories at Lindsay, Horwood and Hawthorns Halls require sufficient holding cold storage and reheating appliances to meet the total demand outlined.

It is estimated that the ratio of demand is proportioned:

Horwood Refectory	45%
Lindsay Refectory	30%
Hawthorns Refectory	25%
Total	100%

Refectory services operate between 12.00 hours and 14.00 hours for lunch and 17.00 and 19.30 hours for dinner, Monday to Friday inclusive.

installation of the most suitable equipment available at the time. The team who produced the feasibility study had determined the throughput necessary to provide for adequate supply throughout the year, but it had no means of determining the type of or latest developments in blast-freeze tunnels, storage refrigerators, etc. Consequently each contractor was supplied with copies of:

(a)   Detailed specification (Table 1).
(b)   Details of annual meal consumption in the three refectories.
(c)   An analysis of the existing five-week menu cycle.
(d)   The choice of dishes available in the five-week menu cycle (Table 2).
(e)   An estimate of the annual consumption of food.
(f)   A plan of the refectory kitchen intended for use as a central production unit.

As part of the detailed specification, contractors were asked to state clearly how the capacity of the scheme recommended had been determined. The object was to enable the team to compare its calculations with those of the contractor and ensure that adequate allowance for blast-freeze tunnel, cold storage and re-heating oven capacity had been made.

TABLE 2
Choice of Dishes Available in Original Five-week Menu Cycle

Fish
Fillet of Fish Meunière
Fillet of Fish Duglère
Fillet of Fish Bretonne
Fillet of Fish Bonne Femme
Fillet of Fish Mornay
Fillet of Fish and Tomato Sauce
Fillet of Fish and Parsley Sauce
Fillet of Fish and Tartare Sauce
Fillet of Fish in Breadcrumbs
Prawn Patties
Fish Cakes

Roasts
Roast Loin of Pork
Roast Leg of Pork
Roast Beef
Roast Lamb
Roast Chicken

Braised and Grilled Liver
Pork Chop

Grills
Gammon and Pineapple
Mixed Grill

Entrees
Braised Steak
Curry and Rice
Cottage Pie
Cornish Pasty
Chicken and Mushroom Vol-au-Vent
Chinese Pork and Rice
Curried Eggs and Rice
Hamburger and Onions
Macaroni au Gratin
Meat Croquettes
Navarin of Lamb
Panhaggerty

TABLE 2—*contd.*

Risotto
Ravioli au Gratin
Scotch Eggs
Sausage Toad-in-the-Hole
Steak, Kidney and Mushroom Pie
Steak and Kidney Pie
Sausage Rolls
Spanish Bake
Sausage Lyonnaise
Spaghetti Bolognaise
Spaghetti Milanaise
Sauté Chicken Chasseur
Savoury Cheese Pudding

Vegetables
Peas
Carrots
Sprouts
Cabbage
Runner Beans
Baked Beans
Swedes
Spring Greens
Parsnips
Cauliflower

Potatoes
Roast
Creamed
Parsley
Sauté
Lyonnaise
Boiled

Sauces
Apple
Stuffing
Onion
Bread
Piquant
Parsley
Sharp
Tomato

Sweets
Apple Charlotte and Custard
Apple Tart and Custard
Apple and Blackberry Tart and Custard
Apple and Loganberry Pie and Custard
Apricot Crumble and Custard
Apricot Tart and Custard
Banana Custard
Bread and Butter Pudding
Bakewell Tart and Custard
Baked Plum Sponge and Custard
Cream Caramel
Cherry Tart and Custard
Cabinet Pudding
Eve's Pudding and Custard
Fruit Jelly and Ice Cream
Ice Cream Sundae
Ice Cream and Chocolate Sauce
Ice Cream and Butterscotch Sauce
Lemon and Orange Chiffon
Lemon Meringue Pie
Mincemeat Tart and Ice Cream
Orange Meringue Pie
Pineapple Upside down and Custard
Peach Flan and Cream
Plum Tart and Custard
Pear Helene
Pineapple Flan and Cream
Queen of Puddings
Rice Pudding
Rhubarb Tart and Custard
Rhubarb and Ginger Crumble and Custard
Sago Pudding
Sherry Trifle and Cream
Strawberry Chiffon
Sliced Peaches and Custard
Strawberry Whip
Steamed Marmalade Sponge and Custard
Steamed Fruit Sponge and Custard
Steamed Chocolate Sponge and Vanilla
　　Sauce
Steamed Apple Dumpling and Custard
Steamed Syrup Sponge and Custard
Treacle Tart and Custard

Evaluation of the Tender

Three tenders were returned and opened on 7 March 1972. It proved impossible to make immediate valid assessment of the tenders submitted and a decision was deferred until a detailed comparison had been made.

Of the three tenders submitted one was discarded on the grounds of substantial price difference and because it was based on a variable price whereas the specification had asked for a fixed one. Analysis of the two remaining tenders showed that, apart from the capacity of the convector ovens, very similar equipment manufacturer and specifications were recommended. However, the major factors influencing the choice of contractor were found to be the greater capacity of blast-freeze tunnel, central cold store and peripheral storage, although before the tender was accepted some adjustments were necessary in the capacity of convector ovens recommended.

It was not until 30 March 1972, that the contract was finally awarded to Staffordshire Refrigeration and Air Conditioning Limited, Stoke-on-Trent, at an agreed price of £34 275·99. The chosen design was prepared under the overall direction of Tom Simpson by Cecil Machin utilising refrigeration equipment manufactured by Fosters Refrigeration (UK) Ltd.

Completion date was agreed as 17 July 1972.

INSTALLATION PROGRAMME

A programme of installation of implementation was agreed:

17 July 1972	Equipment installed.
14 August 1972	Conversion equipment and installation including experiment and testing of recipes (four week period).
11 September 1972	Commence introduction to servery at Lindsay Hall.
9 October 1972	Commence introduction to servery at Hawthorns Hall.
13 November 1972	Commence introduction to servery at Horwood Hall (includes special functions).
11 January 1973	Fully operational at commencement of second term.
January–February 1973	Testing and improving recipes for special functions and conferences.

February–March 1973   Kitchen production and financial control
                       finalised.
April 1973   Assessment.

INITIAL ASSESSMENT FOR THE PERIOD TO 31 DECEMBER 1973

At the end of the financial year in July 1973, the catering establishment was 106 and the target of 104 was reached in September 1973. The transfer of an Assistant Catering Officer to the vacant post of Assistant Domestic Bursar in October 1973, further reduced the establishment.

Additionally, the unexpected saving in food raw materials costs helped to achieve greater overall saving than originally anticipated.

During the first ten months of full implementation to 31 December 1973, the actual financial savings achieved were £10 645, representing £12 775 in a full financial year.

REFECTORY SERVICES FINAL WORK STUDY REPORT

The Cook–freeze Scheme Implementation and Assessment

At the meeting of the Halls of Residence Committee held on 9 December 1971, approval was given to install the Cook-freeze Scheme in the refectories subject to the Medical Officer of Health's approval. The Medical Officer of Health actively encouraged the adoption of the proposed scheme, which, in his opinion, reduced the health hazards of mass catering.

The report covers the period of implementation of the cook–freeze scheme and gives an assessment of the whole catering activity as at 31 March 1973. Further recommendations on the use of the catering resources available resulting from the cook–freeze scheme are also outlined for consideration.

The report is divided into seven sections.

1. Background Information

1.1 In January 1972 the full specification and conditions of tender were compiled and offered to three reputable Companies who were experienced in installing this type of equipment.

1.2 The Tenders Committee received and opened on 7 March 1972 three invited tenders. After careful examination, the contract was awarded to Staffordshire Refrigeration and Air Conditioning Limited, Stoke-on-Trent at a fixed price of £34 275 on 30 March 1972.

1.3 The importance of choosing the right site for the central kitchen production unit cannot be over emphasised. Any unnecessary movement of the food only adds to the operating costs which reduces the effect of the financial savings without adding value to the product. The size and location of the central production unit should be determined by:

(a) The geographical location and distance between the refectories; and
(b) The total amount of meals to be supplied bearing in mind fluctuations and/or peak periods.

Using this information one of the following three types of central production unit may be chosen:

(a) *A new central production installation.* A new installation can incorporate sophisticated equipment with separate food lines for individual meal items to give large-scale output similar to a factory operation.
(b) *A major modification to an existing kitchen.* A smaller-scale operation might combine and re-group existing with new equipment to give a positive production line flow.
(c) *A minor modification to an existing kitchen.* This option might only add sealing, portioning, containering, blast-freezing and central cold-storage facilities.

Option (b) was selected as most appropriate to Keele University's resources and requirements.

1.4 The siting of the cook–freeze production unit at the Lindsay Refectory was due mainly to four reasons which the other two refectories could not match, namely:

(a) *Location:* Lindsay Refectory is geographically most central and about $\frac{1}{4}$ mile from Horwood Refectory whilst Hawthorns Refectory is $\frac{3}{4}$ mile away in the opposite direction.

    (b) *Kitchen capacity:* Total kitchen area is 1020 sq. ft, rectangular in shape and about 34 ft long and 30 ft wide. The kitchen equipment provided by the Universities Grants Committee is capable of supplying 800 meals in two sittings for breakfast, lunch and dinner each day.

    (c) *Food storage:* Central food store for dry goods; all meats and the butchery were already situated there.

    (d) *Central food cold-storage requirement:* Provided by converting the pastry preparation room and one of the dry-goods stores.

1.5   The preparatory work of scheduling and planning the whole cook–freeze system was finalised in accordance with an agreed programme.

1.6   The Senior Assistant Catering Officer was appointed Production Manager by the Catering Officer and the collection of essential information and training of staff in the new technique commenced.

1.7   The central bulk food stores was reorganised and redistributed to allow certain areas to be cleared for the location of freezer tunnel and central cold storage requirements.

1.8   Visits were made to the ICI Central Food Production Kitchen in Ardeer, Scotland, to study production methods and recipe techniques. This was particularly useful as Keele uses the ICI recipe manual for its conventional dishes and sweets. The visit gave the Production Manager a clearer understanding of his responsibilities in carrying out his new role at Keele.

1.9   At another ICI factory the servery service and staffing arrangements of several dining rooms were studied by the Catering Officer and Supervisor in order to balance oven usage with customer demand.

2. Production Kitchen

2.1   Recipe testing in the production kitchen commenced once the cook–freeze equipment was commissioned on 17 July 1972.

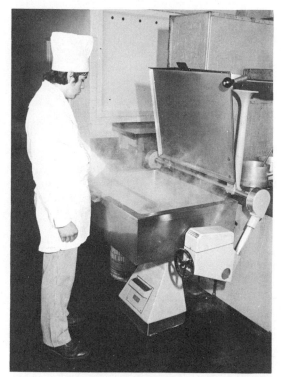

FIG. 1. Primary cooking in the central production kitchen.

2.2 Initially the Catering Officer instructed the Production Manager to carry out the recipe testing personally with the assistance of two cooks.

2.3 The production team was chosen and trained in the new techniques to produce the highest standard and quality of dishes using the best possible quality of food ingredients.

2.4 The controlled quality of food products enabled each dining room to have a common standard rather than under the conventional cooking arrangements being dependent upon six individual kitchen teams which were difficult to maintain and supervise.

2.5 The present strength of the production team consists of: one chef, one assistant cook and four kitchen assistants/packers. When all

FIG. 2.   Frozen foods entering forced air convection oven for reheating.

the food experiments and research on the new range of dishes has been performed one of the kitchen assistants will be withdrawn.

2.6   Initial recipe testing of all main course dishes and sweets was centred on converting the dishes which were produced satisfactorily in accordance with the five-week menu cycle. Each dish variety was batched to produce 100 portions to enable the recipe amounts and costs to be easily calculated by moving the decimal point.

2.7   The combined methods and standard times for each tested and approved recipe were recorded to establish the kitchen equipment usage and working space area required. This enables the production team to operate efficiently and economically. Further basic information collected for every individual dish in the food pack containers included:

(a) Number of food portions and weight of pack.
(b) Freezing time and weight of frozen food pack.
(c) Oven reheating setting-temperature and time taken to cook dish.
(d) Whether lid of food pack remains on or is removed for cooking.

2.8 A revision of stock control procedure and issuing system to the three refectories ascertained the total food portions required for a four-week period to enable the daily kitchen production to be planned accordingly for the full 20-day work cycle.

2.9 The production team of five works a normal eight-hour day and produces on average 4500 portions of food daily. The signatures of both production manager and store-keeper confirm the food portions achieved.

3. Changes in Kitchen Techniques

3.1 Cooking conventionally, the activity in the kitchen builds up to a climax as the meal time approaches and is maintained during service.

3.2 Kitchen area and equipment is now shared between the cook–freeze production and acts as a finishing kitchen to Lindsay serveries. This is not ideal, but, works better than anticipated. By careful planning it was possible to remove several steamers and boilers to create more working surface area in the new production kitchen.

3.3 The principles of flow production were adopted and used separately for both main course dishes and sweets.

3.4 Appendix A shows the adopted recipes quantified and methods used for braised steak with portion costs of raw material ingredients only.

3.5 As every dish is produced in accordance with this pattern of procedure, the following details will clearly show the precise control and quality standards which apply through each step of production.

Fig. 3.   Part of the central cold store holding 17 tons of food.

3.6   The Production Manager had previously requisitioned all the food ingredients to be available for the day's production. Appendix B gives details of the actual daily kitchen production record. The production team is given a copy of the day's production and each recipe and method they must use for each dish (see Appendix A).

3.7   The butcher had previously divided the 125 lb of braising beef required into 500 individual steaks of 4 ozs each. The recipe used was tested and approved on 29 August 1972 and in the event of being revised or changed in any way the new date would be substituted.

3.8   The kitchen assistants prepare the vegetables whilst the chef fries the steak. The carrots are browned and the onions sweated and the liquid is thickened with flour.

3.9   Sauces containing conventional flour are difficult to freeze and reconstitute without breaking down the consistency of the sauce. A means of overcoming this problem is the use of a waxy maize flour; hence 'freeze 'n' flow' is used which cannot be distinguished in the finished product from the conventional.

3.10   Steaks are braised in the usual way and when cooked are taken from the liquid and placed, ten at a time, in aluminium foil containers, with two pints of liquor added. The packs are sealed with lids which have previously been prepared and placed on trolleys and left in the freeze tunnel for 90 minutes.

3.11   The normal freezing time depends upon the contents, density and thickness of the food and can vary from 1 to 2 hours. The equipment has been adjusted to ensure all dishes are frozen in $1\frac{1}{2}$ hours.

3.12   Appendix C shows diagrammatically the procedure for cook–freeze processing of all dish and sweet courses. Certain pastry dishes and sweets are either partially cooked or remain raw and uncooked when placed on a trolley ready for freezing.

3.13   Appendix D lists the standardised range and types of aluminium foil containers and covers used. Where the deep fried method of cooking is required for breaded and battered fish etc., they are frozen on trays and placed in polythene bags for storage.

3.14   The kitchen area is only 20 ft from the two rooms used for central cold storage and is joined by a passageway 5 ft wide. The freeze tunnel was specially designed to fit an available space 14 ft long and 5 ft wide allowing for four trolleys in the tunnel at one time. Each trolley has 14 shelves and a total capacity of 84 packs with 6 packs per shelf.

3.15   Once the food packs on the trolleys have been put into the freezer tunnel it is then the storeman's responsibility to check and remove the food to the cold rooms and to transfer the frozen food on to the shelves.

FIG. 4. Transport van loaded to deliver packs of frozen food to peripheral kitchens.

3.16 The largest cold room has a capacity of 12 600 packs and is used for the main-course dishes. The other cold room with a capacity of 6 608 packs is for sweet dishes.

3.17 The dish portions required by the refectories follow a normal requisition procedure and the food is distributed from central cold storage to the three refectory outlet points 20 yards, $\frac{1}{4}$ mile and $\frac{3}{4}$ mile away, respectively. The food packs are transferred in wire trays holding 12 to 16 packs depending upon the type of container used with a maximum weight of about 40 lbs.

3.18 It is essential that the food temperature should be kept constant while the food is distributed to peripheral kitchens as fluctuations in temperature may affect both the palatability and texture of the food. The vehicle used for distribution may vary, depending on requirement, from a small van using insulated containers to a fully refrigerated vehicle.

3.19 Special refrigerated boxes for the protection of food whilst in transit have not been necessary as the van loading, travelling and unloading into cold storage in refectories takes no more than fifteen minutes.

4.   Refectory Outlets

4.1   The outlet kitchens each have a special cold room built for their supply of food packs. The size of these cold rooms depends upon the number of meals served by the kitchen and was separately calculated to enable a week's supply of food packs to be stored.

FIG. 5.   Food is taken from a peripheral kitchen's cold store. Each store holds one week's supply of frozen foods.

4.2   The supervisor and cook remove food portions required from the cold rooms. The number and size of convection ovens required for reheating the food are related to the maximum number of meals served in each refectory. The capacity of the reheating ovens allows for their use 2–3 times during each meal.

4.3   The dish varieties do not remain long in either the hot cupboards or bainmaries in the servery where food deterioration and loss of nutritional value can occur. With the exception of preparing soups, eggs, bacon, custards and some bought-in prepared vegetables, no other work is performed in the kitchen, hence the main reduction in kitchen staffing is: cooks, kitchen assistants and porters.

4.4   The finishing kitchen using cook–freeze food has the advantage of minimum staffing, yet able to carry out their duties in peace and quiet in a comparatively relaxed atmosphere.

4.5   The most significant change in the service counter is that the containers of cooked food are bright and attractive, and what is important, the food looks quite appetising.

4.6   Portion control is simplified as there are comparatively few portions in each pack and the staff know from the lid details how many portions they need to obtain from each container.

4.7   The servery supervisor warns the cook before the servery has used up all the food choice in the hot cupboard to allow the additional food to be cooked before the dish choice runs out completely.

4.8   Using the conventional service the food was served from large aluminium containers and was difficult to decorate and to do justice to the food. It often lacked customer appeal and tended to portray the worst kind of institutional catering image.

4.9   The cook–freeze method was introduced to certain conference meals served during the Christmas vacation and further extended to the Easter vacation and will be used for most conference meals in the future.

4.10 Students, staff and conference members have welcomed the change which is creating a better dining room atmosphere. This reflects on the servery staff who have a higher regard for the job they perform and have more pride in carrying it out, which goes some way towards job satisfaction and enrichment.

5. Quality and Portion Cost Comparison

5.1 No completely new dishes were introduced during the first five months of actual operation from September 1972 to January 1973 inclusive. The kitchen production was therefore confined to the 112 tested and approved recipes. Further new dish varieties will be added to the list in Appendix E when the recipes are fully tested and approved. The total choice available could reach 200 by the end of 1973 if the present rate of progress is maintained.

5.2 Should there be any customer complaints, or to check how long the food has been stored, the date when the dish was produced is written on every container lid. Periodic checks are being carried out on all dishes and sweets by the University's Biology Department for food deterioration and unwanted bacteria.

5.3 Where braised steak is supplied by the butcher in individual portions the intended output is more readily achieved than when portioning. Where, for example, inaccuracy occurs in the carving of roast meats, this will affect the number of portions achieved and, in turn, the food costs.

5.4 The unit cost is based on 100 portions and the actual portion cost is determined for food ingredients and raw materials only. Due to the fluctuation in food prices, seasonal adjustments are made regularly.

5.5 Due to the wage and equalisation award granted on 1 April 1973, the 38% is increased to 40% of the raw food value and is added to food portion costs to recover all direct wages of the production team and the overhead element. The overhead element includes the capital recovery of the equipment over 10 years, the actual running

FIG. 6.   Trolley containing packs of food entering the blast freezer tunnel.

costs of power, heating and lighting, etc., the Production Manager's salary and all other items such as containers and lids. The actual portion cost of the finished food product forms the selling price to each refectory and although no financial transfers are made this costing is used when performance is compared.

5.6   A number of dishes produced commercially have been bought. When tested and tasted they have not been up to the quality or standard that one would expect and perhaps have done much to damage the frozen product image in the eyes of the consumer and public in general. By careful selection, several suppliers of cook–freeze dishes and sweets meet our standards and in the near future a business relationship will be established with the firms concerned. These firms quite naturally specialise in producing large quantities of a limited number of main courses and sweets and they only produce about 70 of our varieties or just over half our present total range.

5.7 The precise food portion cost comparison with the commercial equipment is difficult as their recipes are not known, and portion weight and size vary. The 70 varieties available include an element for transport and profit. However, a comparison of 37 varieties of main course dishes and sweets shows a cost of £2·98 compared with £4·15 for similar items produced commercially. This represents 39·3% overall cost reduction.

5.8 A business relationship with the commercial producers is an essential safeguard against equipment failure. Insurance valued at £10 000 has been placed against loss or damage to the finished food products to recover the full costs of replacement of raw foods, wages and overheads.

5.9 The Catering Officer purchases most food items in bulk. No financial saving in the food costs was envisaged. However, with the increased variety of dishes, savings were achieved by the more frequent use of savoury and farinaceous dishes which, although cheaper, are just as wholesome and nutritious as wholly meat dishes. In addition to the saving on meals not taken by the catering staff who are no longer employed, the effect of production control of the central and outlet kitchens has produced a saving in food costs which, if maintained, represents £3000 in a full year, or 3% of the £100 000 food budget. The food costs need to be monitored over a full financial year to ensure consistency and the £3000 will help balance the effects of inflation and food prices in general.

5.10 The University's policy that there would be no redundancy has meant that natural wastage alone has accounted for the reduction in kitchen staff. When the scheme was originally started in January 1972 the staff employed was 121, covering the whole of the catering activity and not just the three refectories. However, staff employed in the Adult Education Servery, Senior Common Room Servery, Linen Room, etc., are not affected by the cook–freeze scheme.

The present establishment is 107 and the target is 104 by the end of the financial year, 31 July 1973.

6.   Assessment of the Cook–freeze Scheme

6.1   The original purpose of introducing the cook–freeze scheme in the refectories was to maintain the quality and variety of choice of the conventional food more economically. The benefits of centralisation of food preparation has achieved a reduction in operating costs and generally improved the efficiency of the catering activity without loss of nutritional value, quality or variety.

6.2   In fact a far wider choice and variety is now available with an improvement in flavour and presentation which is most noticeable and to a large extent not originally envisaged.

6.3   The customer reaction and acceptance is most favourable and may have a direct bearing on the reduced amount of food wastage placed in the swill bins after meals.

6.4   The estimated financial saving of £13 069 a year was mainly on labour or £6727 net. When these figures were calculated in May 1971 the effects of further wage and equal pay awards were not included and it could be claimed that this saving on labour is proportionately greater.

6.5   The unexpected savings on food raw material costs helped to cancel out the three remaining members of staff who have not yet left our employment. Thus the financial aims have been achieved and are likely to be in the region of £10 000 in a full year.

6.6   Appendices F and G show the income and expenditure respectively of the weekly catering account for the 35 weeks to 31 March 1973. The financial analysis of the variable to date shows a £5448 surplus which will assist in offsetting the standard expenses of the Easter closure and the increases in wages and equalisation awards brought forward to 1 April 1973.

6.7 The assessment continuity of the production and financial controls implemented will be provided regularly by the Residential Services Costing Assistant once the Work Study services are withdrawn.

7. Further Recommendations

7.1 The main objectives in introducing cook–freeze systems into refectories having been achieved, consideration should be given to obtaining further economies.

7.2 The central kitchen's level of production is entirely independent from the day-to-day operation even though it is related to the number of meals served in the refectories. The flexibility of the cook–freeze operation will, however, allow production to be for example, doubled without any increase in capital expenditure simply by working two shifts. Even without the increased off-take necessary to justify an increased shift, increased production can be achieved at no cost by simply varying shift patterns. For example, production unit staff can be offered the opportunity of working four ten-hour shifts per week rather than the traditional five eight-hour shifts. On the one hand the staff achieve the benefit of a three-day weekend, whilst on the other management gain by the decrease in waiting and rest times and by longer production runs which both contribute to increased productivity. Production schedules and planning also become considerably simpler.

7.3 In order to maximise the benefits of the cook–freeze operation it is necessary to find outlets that provide sufficient meal demand to justify significantly increased production. It is recommended that the following possibilities be explored:

(a) *Internal:* Only approximately two thirds of Keele students subscribe to an all-in meal system. Every effort should be made to attract the remaining students, either on to the all-in system or to persuade them to purchase more casual meals. The two factors that may be used to provide this attraction are the consistent quality of the meals provided and the highly competitive unit price at which they can be offered.

(b) *External:* Every opportunity should be exploited for the supply on a commercial basis of frozen entrees and sweets to other institutions. With carefully planned production schedules no difficulty should be experienced in offering to supply the required dishes to an exact specification provided by the potential customer, provided that they can guarantee an off-take of sufficient volume.

APPENDIX A: RECIPE, METHOD AND COST FOR BRAISED STEAK

Recipe/Method/Cost
Product: Braised Steak
Test date: Tuesday 29 August 1972
Recipe used: Monday 12 March 1973

Actual Quantities Used	Unit cost	Actual cost
100 × 4 oz slices braising beef	0·36p/lb	9·00
4 lb onions	2·40p/56 lb	0·17
4 lb carrots	0·60p/28 lb	0·09
2 lb dripping, bouquet garni		
20 pints brown stock, seasoning		0·05
9 oz plain flour	1·75p/70 lb	0·01
9 oz waxy maize	4·38p/25 kg	0·04
Total cost		£9·36

£0·094 per portion

Method Used
1. Season steaks with salt and pepper.
2. Clean and slice carrots and onions—can be prepared in advance.
3. Melt dripping in bratt pan until hot. Fry steaks until brown.
4. Remove steaks from pan. Lightly brown carrots and onions in pan.
5. Put steaks on to vegetables. Add stock, bouquet garni and seasoning. Bring to the boil. Skim off fat and scum and simmer for $1\frac{1}{2}$ to 2 hours.
6. When steaks are cooked, mix flour and waxy maize with water to form a paste of pouring consistency.
7. Remove steaks from cooking liquor. Place into foil packs—10 × 4 oz steaks per pack.
8. Add flour and waxy maize to cooking liquor and cook for few minutes. Check colour/consistency/flavour. Add 2 pints or 1 lb of sauce to each foil pack.
9. Lid and freeze for 1 hour 30 minutes.

Number of portions produced by above quantities: 100
Pack used for freezing: $9\frac{1}{2}$ in × $9\frac{1}{2}$ in × $1\frac{1}{2}$ in
Weight per pack: pre-freezing—3 lb 0 oz; frozen—3 lb 1 oz

Time in freezer: 1 hour 30 minutes
Reheating information:

	Large	Small
Convection oven setting	8	8
Time allowed (minutes)	45	40

Results and comments: lid on pack whilst cooking.

APPENDIX B: DAILY CATERING PRODUCTION RECORD

PRODUCTION CODE 1203   DATE: Monday 12 March 1973

Item	Number of portions required	Number of portions achieved	Portions per pack	Cost per portion (p)	Total cost (£)
Braised Steak	500	500	10	9·4	47·00
Roast Leg of Pork	500	490	10	9·6	47·04
Roast Leg of Lamb	500	510	10	9·1	46·41
Chicken and Ham Pie	300	306	6	6·5	19·89
Roast Chicken	400	400	8	8·0	32·00
Sausage Lyonnaise	400	416	8	6·4	26·60
Rhubarb and Ginger Tart	400	408	8	2·4	9·79
Rice Pudding	400	392	8	2·0	7·84
Apricot Tart	400	400	8	2·1	8·40
Apple Charlotte	400	408	8	1·9	7·75
Cherry Tart	400	408	8	2·5	10·20
TOTAL	4 600	4 638			£263·92

SIGNATURE
STOCK RECORD CARD ENTERED BY
STOCK RECORD CARD DATE ENTERED ON

APPENDIX C: PROCEDURE USED FOR COOK–FREEZE PROCESS OF ALL MAIN AND SWEET COURSES

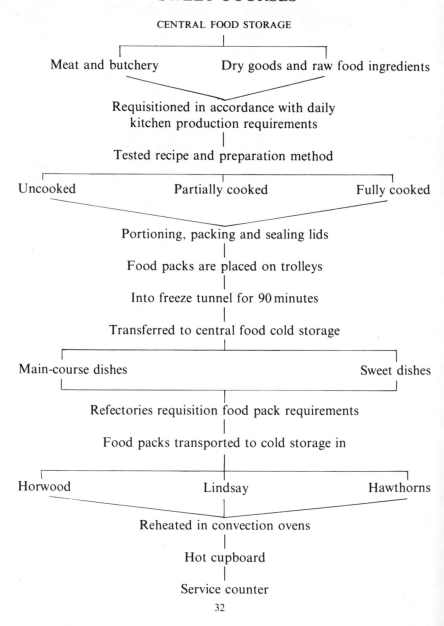

CENTRAL FOOD STORAGE

Meat and butchery | Dry goods and raw food ingredients

Requisitioned in accordance with daily kitchen production requirements

Tested recipe and preparation method

Uncooked | Partially cooked | Fully cooked

Portioning, packing and sealing lids

Food packs are placed on trolleys

Into freeze tunnel for 90 minutes

Transferred to central food cold storage

Main-course dishes | Sweet dishes

Refectories requisition food pack requirements

Food packs transported to cold storage in

Horwood | Lindsay | Hawthorns

Reheated in convection ovens

Hot cupboard

Service counter

APPENDIX D: TYPE OF FOIL CONTAINERS AND COVERS USED

Use	Container	Quantity	Cost (£)	Pack cost (£)
Main course	Deep TP 3636 10973 $9\frac{1}{2}$ in × $9\frac{1}{2}$ in × $2\frac{1}{2}$ in	1 000 Lids 20983	28·86 5·92	0·034 78 set
Main course	Shallow TP 3333 10977/0 $9\frac{1}{2}$ in × $9\frac{1}{2}$ in × $1\frac{1}{2}$ in	1 000 Lids 20983	27·63 5·92	0·033 55 set
Main course	9 in diameter V edge 10819/0 2 in deep	1 000 Lids 20833/0	17·90 5·67	0·023 57 set
Pies/Tarts	9 in diameter Plain 10821/0 1 in deep	1 000 Foil covers 10014	11·89 3·73	0·015 62 set
Puddings	4 in diameter TP 187 UF 2 in deep	1 000 Lids 187 FO	8·71 (includes lids)	0·008 710 each

Use	Container	Quantity	Cost (£)	Pack cost (£)
Puddings	4 in diameter 152/1 RE 045 $1\frac{1}{2}$ in deep	1 000	3·09	0·003 09 each
Individual servings	$4\frac{1}{2}$ in × 3 in TP 1616 10407/0 1 in deep	1 000	7·72	0·007 72 each

APPENDIX E: LIST OF COOK–FREEZE DISHES AND SWEETS AT 31 MARCH 1973

BEEF DISHES
Beef Croquettes
Beef Hot Pot
Braised Liver/Onions
Braised Steak
Cottage Pie
Curried Beef
Meat/Potato Pie
Meat/Potato/Onion Pie
Minced Beef and Onions
Roast Beef
Roast Sirloin of Beef
Sausage Lyonnaise
Sausage Toad-in-the-hole
Sauté Beef and Carrots
Scotch Pie
Steak and Kidney Pie
Steak, Kidney and Mushroom Pie
Steak and Mushroom Pie
Steak and Onion Pie
Steak and Kidney Pudding
Steak, Kidney and Mushroom
   Pudding
Steak and Mushroom Pudding
Steak and Onion Pudding

LAMB DISHES
Cassoulet
Caucasian Shashlik
Grilled Lamb Chop
Lamb Chop Garni
Loin of Lamb Boulangere
Roast Leg of Lamb
Roast Loin of Lamb
Saddle of Lamb
Stuffed Shoulder of Lamb

PORK DISHES
Braised Gammon/Madeira Sauté
Braised Ham Southern Style
Chinese Pork
Pork Croquettes
Grilled Pork Chop
Roast Leg Pork
Roast Loin Pork

FISH DISHES
Fillet Fish Bonne Femme
Fillet Fish Bretonne
Fillet Fish Duglère
Fillet Fish Florentine
Fillet Fish Mornay
Fillet Fish Princesse
Fillet Fish Veronique
Fillet Fish Waleska
Fried Cod in Batter
Fried Fish in Breadcrumbs
Grilled Fish
Scampi Mornay

POULTRY DISHES
Chicken à la King
Chicken Croquettes
Chicken Curry
Chicken and Ham Pie
Chicken and Mushroom Pie
Chicken and Ham Fricassé
Chicken and Mushroom Casserole
Chicken Princesse
Roast Chicken
Sauté Chicken Chasseur
Sauté Chicken Liver
Duckling à l'orange

Roast Duckling
Roast Turkey
Guinea Fowl

FARINACEOUS AND
  SAVOURY DISHES
Lasagne
Macaroni au Gratin
Macaroni au Gratin Savoury
Noodles and Prawns
Panhaggerty
Ravioli au Gratin
Risotto
Spaghetti au Gratin
Spaghetti Bolognaise
Spaghetti Milanaise

SAUCES
Apple
Curry
Orange
Madeira

GRAVIES
Braised Steak Gravy
Lyonnaise Gravy
Pork Gravy
Roast Beef Gravy
Sauté Chicken Gravy
Steak/Kidney Pie Gravy
Steak/Kidney Pudding
Spaghetti Bolognaise
Spaghetti Milanaise
Fish Mornay
Lamb

VEGETABLES
Croquette Potatoes
Dauphine Potatoes
Duchesse Potatoes

Lyonnaise Potatoes
Roast Potatoes
Sauté Potatoes
Cabbage
Cauliflower

COMPLEMENTARY TO MAIN
  DISH
Boiled Rice for Curry
Dumplings
Pie Bases/Pastry Cases
Stuffing Balls
Yorkshire Pudding

SWEET DISHES
Apple Charlotte
Apple Crumble
Apple Dumpling
Apple Eves Pudding
Apple Tart
Apple and Blackberry Tart
Apple and Loganberry Tart
Dutch Apple Tart
Apricot Crumble
Apricot Dumpling
Apricot Tart
Apricot Upside-down
Bakewell Tart
Bread/Butter Pudding
Cabinet Pudding
Cherry Tart
Chocolate Sponge—steamed
Chocolate and Fruit Sponge—
  steamed
Cream Caramel
Damson Tart
Eves Pudding
Fruit Suet Dumpling—steamed
Fruit Sponge
Gooseberry Charlotte

Gooseberry Eves Pudding
Gooseberry Tart
Marmalade Sponge—steamed
Mincemeat Tart
Peach Tart
Pear Upside-down
Pineapple Flan
Pineapple Upside-down
Plum Crumble

Baked Plum Sponge
Plum Tart
Rhubarb Eves Pudding
Rhubarb and Ginger Crumble
Rhubarb and Ginger Tart
Rice Pudding
Sherry Trifle
Syrup Sponge—steamed
Syrup Tart

APPENDIX F:

CATERING COST CONTROL SUMMARY WEEK ENDING SATURDAY:-		TOTAL		HORWOOD		LINDSAY	
INCOME							
CASH	*FOOD*	196.71		91.57		.78	
CREDIT	,,	7,002.04		3,773.49		2,337.73	
WEEKS TOTAL FOOD INCOME	,,	7,198.75		3,865.06		2,338.51	
PREVIOUS WEEKS ACCUMULATIVE TOTAL	,,	155,906.11		73,591.74		43,197.73	
INCOME TO DATE	,,	163,104.86		77,456.80		45,536.24	
CASH	*TOBACCO*	70.05		24.12		—	
PREVIOUS WEEKS ,, ,, ,,		3,755.55		465.93		1,073.02	
CASH TO DATE	,,	3,825.60		490.05		1,073.02	
CASH	*DRINKS*	643.15		232.35		203.85	
CREDIT	,,	295.87		241.25		49.23	
WEEKS TOTAL DRINKS INCOME	,,	939.02		473.60		253.08	
PREVIOUS WEEKS ACCUMULATIVE TOTAL		9,608.49		3,576.04		2,673.54	
INCOME TO DATE		10,547.51		4,049.64		2,926.62	
TOTAL INCOME THIS WEEK FOOD, TOBACCO, DRINKS		8,207.82		4,362.78		2,591.59	
TOTAL INCOME TO DATE FOOD, TOBACCO, DRINKS		177,477.97		81,996.49		49,535.88	
VARIABLE Plus (RED FOR) THIS WEEK		2,473.17		1,646.17		823.57	
Minus (MINUS) TO DATE		5,448.79		15,343.92		−11,953.31	

	% THIS WEEK	% TO DATE	% THIS WEEK	% TO DATE	% THIS WEEK	% TO DATE
WAGES PERCENTAGE TO INCOME	32.10	35.18	25.68	25.32	34.14	49.14
FOOD ,, ,, ,,	30.76	40.49	27.20	33.17	26.29	55.39
TOBACCO ,, ,, ,,	91.66	91.67	91.66	91.64	—	91.66
DRINKS ,, ,, ,,	19.96	57.65	27.05	55.13	15.26	59.72

SATURDAY   31st MARCH 1973

HAWTHORNS	PRODUCTION UNIT	CHANCELLOR'S BLD.	S.C.R. HORWOOD	VENDING	MISC. & SUNDRIES
NIL	—	48.78	21.43	5.95	28.20
879.87	—	—	10.95	—	—
879.87	—	48.78	32.38	5.95	28.20
34,101.54	—	3,295.97	711.39	940.71	67.03
34,981.41	—	3,344.75	743.77	946.66	95.23
—	—	40.94	4.99	—	—
797.74	—	1,176.25	242.61	—	—
797.74	—	1,217.19	247.60	—	—
206.95	—	—	—	—	—
5.39	—	—	—	—	—
212.34	—	—	—	—	—
3,358.91	—	—	—	—	—
3,571.25	—	—	—	—	—
1,092.21	—	89.72	37.37	5.95	28.20
39,350.40	—	4,561.94	991.37	946.66	95.23
64.98	—	−75.36	12.61	−22.20	23.40
1,852.53	—	203.71	−138.02	650.33	−510.37

% THIS WEEK	% TO DATE	% THIS WEEK	% TO DATE	% THIS WEEK	% TO DATE	% THIS WEEK	% TO DATE	% THIS WEEK	% TO DATE	% THIS WEEK	% TO DATE
49.49	36.43	—	—	139.44	58.98	49.69	91.59	—	—	—	—
—	36.38			—	37.95	—	29.73	—	31.30		6.35
—	91.72	—	—	91.64	91.67	91.58	91.66	—	—	—	—
9.75	58.80	—	—	—	—	—	—	—	—	—	—

APPENDIX G:

CATERING COST CONTROL SUMMARY WEEK ENDING SATURDAY:-	TOTAL	HORWOOD	LINDSAY
EXPENDITURE			
FOOD	2,214.93	1,051.43	614.91
,, PREVIOUS WEEKS ACCUMULATIVE TOTAL	63,831.41	24,646.60	24,611.54
,, ACCUMULATIVE TOTAL TO DATE	66,046.34	25,698.03	25,226.45
WAGES	2,311.03	992.93	798.48
,, PREVIOUS WEEKS ACCUMULATIVE TOTAL	55,078.72	18,619.26	21,580.96
,, ACCUMULATIVE TOTAL TO DATE	57,389.75	19,612.19	22,379.44
ACCUMULATIVE TOTAL TO DATE FOOD & WAGES	123,436.09	45,310.22	47,605.89
INVOICED THIS WEEK *TOBACCO*	64.21	22.11	—
PREVIOUS WEEKS ACCUMULATIVE TOTAL	3,443.04	427.00	983.61
TOTAL TO DATE	3,507.25	449.11	983.61
INVOICED THIS WEEK *DRINKS*	187.48	128.14	38.63
PREVIOUS WEEKS ACCUMULATIVE TOTAL	5,893.29	2,104.65	1,709.32
TOTAL TO DATE	6,080.77	2,232.79	1,747.95
EXPENDITURE TOTAL FOOD, WAGES, TOBACCO, DRINKS	4,777.65	2,194.61	1,452.02
PREVIOUS WEEKS ACCUMULATIVE TOTAL	128,246.46	45,797.51	48,885.43
TOTAL EXPENDITURE TO DATE	133,024.11	47,992.12	50,337.45
TRADING CONTRIBUTION TO OVERHEADS	3,430.17	2,168.17	1,139.57
DIRECT ,, ,, ,,	—	—	—
TOTAL ,, ,, ,,	3,430.17	2,168.17	1,139.57
PREVIOUS WEEKS ACCUMULATIVE TOTAL	43,749.62	32,930.63	− 1,154.96
TOTAL CONTRIBUTION TO OVERHEADS THIS WEEK	47,179.79	35,098.80	− 15.39
STANDARD OVERHEADS THIS WEEK	957.00	522.00	316.00
TOTAL ACCUM. STANDARD OVERHEADS TO DATE	41,731.00	19,754.88	11,937.92

SATURDAY 31st MARCH 1973

HAWTHORNS	PRODUCTION UNIT	CHANCELLOR'S BLD.	S.C.R. HORWOOD	VENDING	MISC. & SUNDRIES
452.01	(370.20)	59.53	4.10	28.15	4.80
12,277.36	(13,146.74)	1,209.92	217.03	268.16	600.80
|12,729.37	(13,516.94)	1,269.45	221.13	296.31	605.60
435.51	(217.50)	68.02	16.09		
12,308.34	(3,663.50)	1,904.96	665.20		
12,743.85	(3,881.00)	1,972.98	681.29		
25,473.22	(17,397.94)	3,242.43	902.42	296.31	605.60
—	—	37.53	4.57	—	—
731.76	—	1,078.27	222.40		
731.76	—	1,115.80	226.97		
20.71	—	—	—	—	—
2,079.32	—	—	—	—	—
2,100.03	—	—	—	—	—
908.23		165.08	24.76	28.15	4.80
27,396.78	—	4,193.15	1,104.63	268.16	600.80
28,305.01	—	4,358.23	1,129.39	296.31	605.60
183.98	—	−75.36	12.61	−22.20	23.40
—					
183.98	—	−75.36	12.61	−22.20	23.40
11,706.75	—	279.07	−150.63	672.53	−533.77
11,890.73	—	203.71	−138.02	650.33	−510.37
119.00	—				
10,038.20	—				

PART II

Recipes

INTRODUCTORY REMARKS

Much has been written about the clear economical benefits of centralised systems catering using blast-freezing techniques. Whilst all this literature is of obvious interest to management and accountants it is of little practical value to the Production Manager and his staff who have the responsibility for producing the food to the satisfaction of the diner and developing the necessary standard recipes. As the first university in Europe to convert to cook–freeze and with only a few experienced operators prepared to give advice, the construction of a standard recipe file proved to be a long and exhaustive task based on trial and error and requiring considerable initiative and patience. This task began in 1968, well before the installation of the equipment, and continues to this day.

The principal differences between conventional cooking and the production of pre-cooked frozen food are selection of ingredients, degree of cooking and consistency of product prior to freezing. As the process of cook–freezing can cause damage to the texture of certain dishes and affect the flavour and taste of others it is essential to modify standard recipes and cooking procedures so that cook–frozen dishes will not suffer by comparison with conventionally prepared dishes. It may also be necessary especially when semi-skilled staff are used to reheat the foods, to adjust the weight of certain dishes so that various types of dish may be placed in the reheat oven at the same time and take the same time to be ready for service.

CHOICE OF INGREDIENTS

General

It is worth remembering, as a general rule, that although the freezing process may not cause many foods to deteriorate there are no foods that it will actually improve. In practical terms this means that any foods purchased should be of the best possible quality, always, of course, commensurate with the requirement for sensible economy. This rule particularly applies to foods, such as meat, which have a high fat content, where freezing may result in changes in taste due to rancidity resulting from oxidation of the fat molecules taking place during storage. To eliminate this problem it is advisable to:

(a)   use lean meat for any dishes containing mincemeat;

(b)   trim away all unnecessary fat so that when the dish is cooked only a minimal layer of fat remains; and

(c)   skim all fat from meat extracts.

Flour- and Starch-thickened Dishes

Substitution of some wheat flour by starch with a high amylopectin content is necessary in order to obtain a satisfactory product in the case of sauces, thickened soups, meat dishes in sauce, and, for example, chicken in sauce. Dishes thickened with wheat starch tend to breakdown and curdle after freezing and reheating. But if dishes are made with starches containing a high proportion of amylopectin (waxy rice, waxy maize and tapioca starches) they will not curdle. Milk puddings and the normal range of cornflour desserts and sauces should be prepared to a slightly thinner consistency than would be normal because starch has a tendency to thicken at lower temperatures and naturally a certain amount of evaporation occurs during reheating. In richer starch-thickened desserts, eggs may be used in addition to the thickening agent.

Gelatin Dishes

Freezing has a detrimental effect on the texture of desserts with a jelly base as the smooth texture is lost and replaced by a granular appearance.

Vegetables

Vegetables should not be fully cooked before freezing. Whilst the actual cooking time will be different for each type and size of vegetable the cooking time must be sufficient to ensure that enzymes are inactivated. For each portion of vegetables 5 ml of water should be added during packing to provide a little steam during reheating.

Changes in Appearance

The colour of vegetables tends to fade due to hydrolysis when storage is at temperatures of higher than $-18\,^{\circ}C$ or even if they are stored for lengthy periods at even lower temperatures. Conversely certain vegetables such as Brussels sprouts that are not sufficiently blanched may change colour and become pink in the middle. Cauliflower changes colour from its normal white to yellow and then from yellow to brown if stored at above −18 °C. The colour of meat tends to fade when it is stored at low temperatures whilst, on the other hand, in poultry there is a noticeable darkening of both the bones and the meat near the bones when it is cooked. In fruit the browning of cut raw surfaces takes place due to the action of the enzyme phenolase. This reaction can be prevented, particularly with apples, apricots, peaches, bananas and avocados by freezing the fruit in a thin sugar syrup.

Critical Items

Foods that deteriorate rapidly after cooking must be frozen immediately. Examples of such foods are omelettes, scrambled eggs and welsh rarebit. Caution must also be exercised in order to ensure that foods are not kept tepid for long periods as these offer an excellent growth medium for micro-organisms.

DEGREE OF COOKING

The cooking times of certain dishes can be reduced as the reheating process obviously partly continues the cooking. Examples of dishes that require a reduced initial cooking time are those dishes with pastry tops such as pies, fried or poached fish and most potato dishes. Cooking time can also be reduced for most vegetables because freezing, in addition to the cooking process, also causes some softening of the product. Well-cooked frozen green vegetables and carrots may have an unpleasant slimy texture after reheating.

PORTIONING

Freezing times depend on the pack content, the thickness and the density of the food. Thus if only semi-skilled staff are employed on the reheating process and as a result it is considered sensible to have standard cooking times for each type of pack, it is clearly necessary to vary the number of portions in packs containing foods of different densities. It is also necessary to strictly control the size of individual items which, if too large, would not reheat satisfactorily in the time allocated. For example, creamed potatoes may be hollowed out slightly at the centre of the tray, as this is the slowest reheating area, so that the whole dish reheats at the same temperature in the same time.

STANDARD RECIPES

A standard recipe is one that has been thoroughly tested and has consistently produced the required result. Standardisation ensures that food to the laid-down standard can be served all the time. Recipes are exact and accurate and all guesswork is eliminated; absent key workers can be replaced by others. The problems of over- or under-production are eliminated as quantities are exact and yields predictable and in consequence food costs can even be determined, if necessary, before the cooking process.

It is important to ensure that recipes are kept under continuous review in order to further improve quality, customer satisfaction and to make work procedures even more efficient. A standard recipe form should be used and should include most of the following information:

 Name of recipe
 Menu file code number
 Equipment to be used
 Purchasing requirements
 Prepared raw materials
 Ingredients in order of use
 Method
 Cooking temperature
 Cooking time
 Yield of the raw mixture
 Total yield of finished product
 Portion size
 Number of portions
 Portion price
 Production time
 Reheating instructions
 Test dates.

KEELE RECIPES

Although only about 90 recipes are included in this publication the Production Manager at Keele has at his disposal some 250 well tried and proven recipes. This large repertoire is necessary in order to sustain not only the normal menu cycle for the term but also to provide dishes for special functions throughout the year. While most of the dishes included on the second list are obviously not cook–frozen, all fish dishes, roasts, entrees and sweets are and these dishes together with those on the day-to-day menus give indications of the range and variety of dishes that can be prepared and presented to a high standard of quality.

One other point worth making is that brief examination of these menus will show that although most students are basically conservative in their feeding habits, some dishes quite unexpectedly achieve a very high popularity rating. This more adventurous attitude is nowhere more noticeable than in the rapid growth in popularity of vegetarian foods. Keele operates a separate vegetarian menu at all meals and examples of popular dishes within this category can be found in the farinaceous section.

Meat

BEEF HOT POT (300 PORTIONS)

60 lb (28 kg)	Cubed beef	1. Season the meat and seal it quickly in hot dripping in a bratt pan.
3 lb (1·5 kg)	Dripping	
	Seasoning	
12 lb (6 kg)	Sliced onions	2. Add the sliced vegetables, herbs and a little more seasoning. Fry until the vegetables are golden brown. Sprinkle with flour, stir well.
12 lb (6 kg)	Sliced carrots	
8 lb (4 kg)	Sliced celery	
	Herbs to taste	
12 oz (350 g)	Flour	
1½ lb (700 g)	Tomato puree	3. Add the tomato puree and stock. Bring to the boil and reduce heat to a gentle simmer. Skim off any excess fat and any scum that rises to the surface. Simmer gently until the meat is almost cooked. Stir occasionally. Add the monosodium glutamate.
60 pints (35 litres)	Brown stock	
1 oz (25 g)	Monosodium glutamate	
18 lb (8 kg)	Sliced potatoes	4. Slice the potatoes about $\frac{1}{5}$ in thick. Parboil in boiling salted water; drain and refresh.
18 lb (8 kg)	Diced potatoes	5. Add the diced potatoes to the meat in the bratt pan; continue to cook gently until the potatoes are almost cooked. Add more stock if and when necessary.

6. Check for seasoning, colour and consistency.

7. Weigh 3 lb (1·5 kg) of hot pot into each 9½ in sq shallow container.

8. Arrange 10 oz (250 g) sliced potatoes on the top.

9. Lid, seal and put into freezer for at least 90 minutes.

10. Remove from freezer and store in holding fridge until required.

Lid Notes
Description: Beef Hot Pot
No. of portions: 8
Oven setting: 6
Oven time: 45–50 minutes
Lid off
Pack weight: 1·75 kg

BEEF STROGANOFF (72 PORTIONS)

20 lb (10 kg)	Fillet of Beef Seasoning	1. Cut the fillet into strips and season.
10 lb (5 kg) 5 lb (2·5 kg) ½ lb (250 g)	Chopped onions Sliced mushrooms Butter	2. Simmer the onions and mushrooms in butter until almost cooked.
2 gals. (9 litres)	Demi-glace	3. Add the demi-glace and finish cooking.
¾ lb (350 g)	Butter	4. Heat the butter in a stock pot lid. Add the seasoned beef a little at a time so that it browns immediately. Turn it over and brown quickly on all sides, leaving the centre of the meat pink. Place the meat on a wire rack to drain.
3 pints (2 litres)	Cream	5. Reduce the liquor left in pan and add the cream. Add the demi-glace, onions and mushrooms and bring to the boil.
2 oz (50 g) 3	Seasoning Made mustard Lemons (juice from)	6. Season the sauce, add a little made mustard and the lemon juice. Check for seasoning, colour and consistency. Add the blood which has drained from the cooked fillet. Toss in the meat.
		7. Put 3½ lb (1·75 kg) of stroganoff into each 9½ in sq deep container.
		8. Lid and seal.
		9. Put into freezer for 90 minutes.
		10. Store in holding fridge until required.

Lid Notes
Description: Beef Stroganoff
No. of portions: 6
Oven setting: 5
Oven time: 45 minutes
Lid on
Pack weight: 1·75 kg

BRAISED LIVER AND ONIONS (100 PORTIONS)

25 lb (12 kg)	Sliced liver (4 oz per slice)	1. Wash and drain the sliced liver.
2 lb (1 kg)	Dripping	2. Heat the dripping in a bratt pan and seal the liver quickly on each side. Remove the liver and put aside for a while.
5 lb (2·5 kg)	Sliced onions	3. Sieve the fat from the bratt pan and fry the onions in it.
28 pints (12 litres) 4 oz (120 g)	Brown stock Tomato puree Seasoning Mixed herbs	4. Add the stock to the onions, add the liver, tomato puree, seasoning, a sprinkling of mixed herbs and chopped celery. Bring to the boil, reduce heat and
4 × A2½ or 3	Heads celery	simmer gently until the liver is cooked. Remove scum and excess fat as it rises to the surface.
		5. Remove the liver from the stock and pack 10 slices into each 9½ in sq deep foil container.
		6. Bring the stock back to the boil.
10 oz (250 g) 1 lb (450 g)	Plain flour Freeze 'n' flow starch Water to mix	7. Mix the flour and starch to a smooth paste with cold water. Stir into the stock. Bring back to the boil, reduce the heat and simmer gently for a few minutes to cook the flour.
		8. Check for colour, consistency and seasoning.
		9. Add 2 pints (1 litre) gravy to each pack.
		10. Lid, seal and put into freezer for 90 minutes.

Lid Notes
Description: Braised Liver and Onions
Oven setting: 6
Oven time: 50–60 minutes
No. of portions: 10
Lid on
Pack weight: 1·8 kg

BRAISED STEAK (100 PORTIONS)

25 lb (12 kg)	Thick flank in 4 oz slices	1. Season the steaks and seal them in hot dripping in a bratt pan. Put them on one side until stage 3.
2 lb (1 kg)	Dripping	
	Seasoning	
8 lb (4 kg)	Sliced carrots	2. Lightly brown the vegetables in the bratt pan.
5 lb (2·5 kg)	Sliced onions	
3 heads (2 kg)	Celery	

3. Put the steaks on to the vegetables.

28 pints (12 litres)	Brown stock

4. Add the stock, tomato puree and herbs. Bring to the boil, reduce the heat to a gentle simmer. Skim off any excess fat and any scum that rises to the surface. Cook gently until the steaks are tender.

16 oz (450 g)	Freeze 'n' flow starch
10 oz (250 g)	Plain flour

5. Mix the starch and flour to a smooth paste of pouring consistency.

6. Remove the steaks from the cooking liquor, take to packing station and pack them 10 to each 9½ in sq deep foil container.

7. Stir the flour paste into the cooking liquor. Bring back to the boil, stirring all the time. Reduce to a simmer and cook the flour. Skim if necessary. Check for colour, flavour and consistency.

8. Put 2 pints (1 litre) sauce on to each pack.

9. Lid, seal and freeze for at least 90 minutes.

10. Store in holding fridge until required.

Lid Notes
Description: Braised Steak
No. of portions: 10
Oven setting: 6
Oven time: 50–60 minutes
Lid on
Pack weight: 1·85 kg

BREADED LAMB CHOP (120 PORTIONS)

120 × 4 oz (120 g)   Lamb chops
2 lb (1 kg)   Flour
   Pepper and salt

6 pints (3 litres)   Egg wash

10 lb (5 kg)   Breadcrumbs

1. Season the flour. Pass the chops through the flour. Shake off the excess.

2. Pass the floured chops through the egg wash. Make sure that they are covered. Drain off the excess egg wash.

3. Pass the chops through the breadcrumbs. Again, make sure that they are covered. Pat them with a palette knife to aid adhesion. Shake off the excess breadcrumbs.

4. Wrap the chops individually in cling film.

5. Put into racks on the freezing trolley.

6. Put into freezer for 90 minutes.

7. Pack in suitable boxes. Label them and put the count on the boxes.

8. Store in holding fridges until required.

Lid Notes
Description: Breaded Lamb Chop
Reheating: Deep fry

CARBONNADE OF BEEF (240 PORTIONS)

60 lb (28 kg)	Cubed beef
3 lb (1·5 kg)	Dripping
	Seasoning

1. Melt the dripping and heat in a bratt pan. Toss the beef in the hot dripping until it is brown. Season well. Remove the beef from the pan, but retain the dripping.

12 lb (6 kg)	Finely chopped onions
1½ lb (650 g)	Flour
7 gals. (32 litres)	Brown stock
4 pints (2 litres)	Beer
2 lb 2 oz (1 kg)	Demerara sugar

2. Quickly toss the onions in the dripping to moisten them. Put the beef back into the bratt pan, stir in the flour. Gradually add the brown stock stirring all the time. Stir in the beer and brown sugar. Test and correct for seasoning.

3. Bring to the boil, reduce heat and simmer until the beef is cooked. Remove excess fat and scum as it rises.

1½ lb (650 g)	Flour
3 lb (1·25 kg)	Freeze 'n' flow starch
	Water to mix

4. Mix the starch and flour to a smooth thin paste with cold water. Stir into the bratt pan, bring back to the boil, stirring all the time, reduce the heat and simmer for 5 minutes to cook the starch.

	Gravy browning

5. Add gravy browning to give a rich brown colour. Test for colour, flavour and consistency. Correct as necessary.

6. Weigh 3½ lb (1·75 kg) of carbonnade into each 9½ in sq shallow container.

7. Lid and seal.

8. Put into freezer for 90 minutes.

9. Store in cold room until required.

Lid Notes
Description: Carbonnade of Beef
No. of portions: 8
Oven setting: 6
Oven time: 40 minutes
Lid on
Pack weight: 1·75 kg

CASSOULET (350 PORTIONS)

28 lb (12·5 kg)	Haricot beans	1. Cover the beans with cold water and soak overnight.
		2. Blanch the haricot beans and drain.
15 lb (7 kg)	Cubed streaky bacon	3. Blanch the bacon.
6	Cloves crushed garlic Pepper Salt Water to cover	4. Put the haricot beans and blanched bacon with the garlic into a bratt pan and just cover with water. Season, bring to the boil, reduce the heat and simmer with the lid closed for a good hour.
		5. Drain off and save the liquor.
30 lb (14 kg) 7 lb (3·5 kg) 1 pint (0·5 litre)	Cubed mutton Cubed duck meat Bouquet garni Seasoning Dripping Tomato puree	6. Fry the mutton and duck meat in good dripping, add the haricot beans, bouquet garni, plenty of pepper and a little salt. Moisten well with the bean liquor and tomato puree and stew slowly for 3–3½ hours. Add a little liquor from time to time.
10 lb (4·5 kg)	Cubed garlic sausage	7. After 2½ hours add the garlic sausage.
3 × A10	Baked beans	8. When the haricot beans are cooked add the baked beans.
		9. Remove the bouquet garni from the beans.
		10. Weigh 3 lb (1·25 kg) of mixture into each 9½ in sq deep container.
30 lb (14 kg) 1 tablespoon	Concassé tomatoes Seasoning Sugar	11. Season the concassé tomatoes and add the sugar.
		12. Spoon 10 oz (250 g) of concassé tomato over the beans etc.
5 lb (2·5 kg)	Fried breadcrumbs	13. Sprinkle with 2 oz (50 g) browned breadcrumbs.
		14. Lid and seal.
		15. Put into freezer for 90 minutes.
		16. Store in holding fridge until required.

Lid Notes
Description: Cassoulet
No. of portions: 8
Oven setting: 6
Oven time: 50 minutes
Lid off
Pack weight: 1·55 kg

COTTAGE PIE (96 PORTIONS)

25 lb (12 kg)	Minced beef	
6 lb (2·75 kg)	Minced onions	
1 lb (500 g)	Beef dripping	
	Seasoning	

1. Fry off the minced beef and onion in the dripping. Season with pepper and salt. Stir frequently to avoid it catching. Remove excess fat as it appears.

6 pints (3·5 litres)   Brown stock

2. Add sufficient brown stock to moisten; sprinkle lightly with mixed herbs. Bring to the boil, reduce the heat and simmer gently until the meat is cooked. More fat will appear during the cooking process. This must be removed. It may be necessary to add more stock from time to time to keep the meat moist.

4 oz (120 g)   Plain flour
4 oz (120 g)   Freeze 'n' flow starch
    Water to mix

3. Mix the starch and flour to a smooth paste with cold water. Add to the mince. Stir well and allow it to simmer for 5 minutes to cook the flour. Adjust seasoning.

6 lb (2·75 kg)   Potato powder
    Boiling water
8   Eggs
    Seasoning

4. Mix the potato powder and boiling water, according to the manufacturer's instructions on the mixing machine. Add the eggs and season to taste.

5. Weigh 2 lb (1 kg) of minced beef into each 9½ in sq deep container.

6. Pipe on 1 lb (500 g) potato through medium star tube.

7. Lid and seal.

8. Put into freezer for 2 hours.

9. Store in holding fridge until required.

Lid Notes
Description: Cottage Pie
No. of portions: 6
Oven setting: 5
Oven time: 50 minutes
Lid off
Pack weight: 1·5 kg

CURRIED BEEF (250 PORTIONS)

60 lb (28 kg)	Cubed beef
12 lb (6 kg)	Sliced onions
2 lb (1 kg)	Dripping
	Seasoning
3 lb (1·5 kg)	Curry powder
2 lb (1 kg)	Plain flour
1½ lb (700 g)	Tomato puree
60 pints (35 litres)	Brown stock
4	Bayleaves
9 lb (4 kg)	Solid pack apple
6 lb (2·5 kg)	Currants
3 lb (1·5 kg)	Mango chutney
3 lb (1·5 kg)	Marmalade
3 lb (1·5 kg)	Desiccated coconut
2½ lb (1·25 kg)	Freeze 'n' flow starch

1. Season the meat and seal it with the onions in hot dripping in a bratt pan.

2. Stir in the curry powder and cook. Add the flour and dry it out, stirring all the time.

3. Stir in the brown stock and tomato puree, bring to the boil, reduce heat to a gentle simmer. Skim off any excess fat and any scum that rises to the surface. Simmer until the meat is tender. Add bay leaves at this stage.

4. Chop the apple and add to the meat, with the currants, chutney, marmalade and coconut. Again remove any scum that rises to the surface.

5. Mix the starch to a smooth paste with cold water. Stir into the curry. Bring back to the boil and simmer gently for 5 minutes, stirring occasionally. Check colour, seasoning and consistency.

6. Remove from bratt pan and take it to the packing station.

7. Weigh 3½ lb (1·75 kg) into each 9½ in sq shallow foil container.

8. Lid, seal and put into tunnel for at least 90 minutes.

9. Remove and store in holding fridge until required.

Lid Notes
Description: Curried Beef
No. of portions: 8
Oven setting: 6
Oven time: 45–50 minutes
Lid on
Pack weight: 1·75 kg

IRISH STEW (100 PORTIONS)

20 lb (10 kg)	Lamb pieces	1. Blanch the lamb pieces in boiling water for 2 minutes and drain. This seals the meat.
5 lb (2·25 kg) 3 lb (1·35 kg) 2 lb (1 kg) 2½ gals. (12 litres)	Sliced potatoes Sliced onions Sliced celery Bouquet garni Seasoning Water	2. Put the potatoes, onions and celery into a bratt pan. Add the blanched lamb. Cover with water, add the bouquet garni and season. Bring to the boil, reduce the heat and simmer gently until the meat is *almost cooked*. Stir gently to stop the vegetables from catching. Remove excess fat and scum as it appears.
		3. When the lamb is almost cooked remove it from the bratt pan. Strain the cooking liquor into a saucepan. Transfer the lamb to a clean bratt pan and cover with the strained liquor.
10 lb (4·5 kg) 5 lb (2·25 kg)	Sliced potatoes Sliced onions	4. Put the sliced potatoes and onions on to the lamb and finish cooking. Check for seasoning.
		5. Carefully pour the liquor off the meat into a saucepan and adjust the quantity to 2½ gals. (12 litres).
		6. The liquor should be slightly thickened by the potato; if it is not thick enough, correct it with freeze 'n' flow starch.
		7. Weigh 2 lb (1 kg) of lamb and vegetables into each 9½ in sq deep pack.
		8. Add 2 × 15 fl. oz (1 litre) ladles of cooking liquor to each pack.
2 oz (50 g)	Fresh chopped parsley	9. Sprinkle with chopped parsley.
		10. Lid and seal.
		11. Put into freezer for 90 minutes
		12. Store in cold room until required.

Lid Notes
Description: Irish Stew
No. of portions: 8
Oven setting: 6
Oven time: 45 minutes
Lid on
Pack weight: 2 kg

LIVER AND BACON PIE (240 PORTIONS)

12 lb (6 kg)	Bacon pieces	1. Soak the bacon pieces overnight in cold water.
		2. Chop or coarsely mince the bacon and cook in sufficient water to cover in a saucepan.
18 lb (8 kg)	Chicken livers	3. Peel and chop the onions and sweat in the dripping. Add the chicken livers and seal them quickly.
5 lb (2·5 kg)	Onions	
1 lb (500 g)	Dripping	
1 lb (500 g)	Flour	4. Stir in the flour.
4 gals. (18 litres)	Water	5. Add the water, tomato puree and thyme stirring all the time until the sauce comes to the boil. Reduce the heat and simmer gently. Season to taste. Remove scum and excess fat as it appears.
½ lb (250 g)	Tomato puree	
	Pinch of thyme	
	Seasoning	
		6. Drain the bacon and add to the chicken livers.
2 lb (1 kg)	Freeze 'n' flow starch	7. Reconstitute the freeze 'n' flow with cold water and stir into the sauce. Slowly bring back to the boil, reduce the heat and simmer gently until the starch is cooked.
	Gravy browning	8. Check the sauce for colour, flavour and consistency, adjusting if necessary and allow it to cool.
20 lb (10 kg)	Shortcrust pastry	9. Line 9 in diameter pie foils with ½ lb (250 g) pastry.
		10. Add 1¼ lb (550 g) of liver and bacon filling.
		11. Top with 6 oz (200 g) pastry.
½ pint (250 cc)	Egg wash	12. Brush with egg wash.
		13. Cover with suitably labelled foil.
		14. Put into blast freezer for 90 minutes.
		15. Store under refrigeration until required.

Lid Notes
Description: Liver and Bacon Pie
No. of portions: 6
Oven setting: 5
Oven time: 35 minutes
Lid off
Pack weight: 1 kg

MEAT, POTATO AND ONION PIE (520 PORTIONS)

40 lb (18 kg) 10 lb (4·5 kg) 40 pints (24 litres)	Cubed steak Sliced onions Brown stock Seasoning Mixed herbs	1. Put the cubed beef and onions into a bratt pan and cover with brown stock or water (approx. 40 pints). Season and sprinkle with mixed herbs. Bring to the boil, reduce the heat and simmer gently until the meat is almost cooked. Skim off any fat and scum as it rises to the surface.
20 lb (10 kg)	Cubed potatoes	2. Add the potatoes to the meat and cook gently until the potatoes are almost cooked. It may be necessary to add more stock or water at this stage.
12 oz (350 g) 12 oz (350 g)	Freeze 'n' flow starch Plain flour Water to mix	3. Mix the starch and flour to a smooth creamy paste with cold water. Add slowly to the bratt pan stirring continuously. Do not over-thicken. Simmer gently until the flour is cooked.
1 oz (25 g)	Monosodium glutamate	4. Add the monosodium glutamate; check for colour, seasoning and consistency.
60 lb (28 kg)	Shortcrust pastry	5. See method for preparation of short-crust pastry.
		6. Line the 9 in diameter pie foils with ½ lb (250 g) pastry.
		7. Add 1¼ lb (550 g) meat and potato.
½ pint (250 cc)	Egg wash	8. Cover with 6 oz (200 g) pastry and brush with egg wash.
		9. Cover with suitably labelled foil.
		10. Freeze for 90 minutes.
		11. Store in holding fridge until required.

Lid Notes
Description: Meat, Potato and Onion Pie
Oven setting: 4
Oven time: 35 minutes
Cover off
Pack weight: 1 kg

MINCED BEEF AND ONIONS (96 PORTIONS)

25 lb (12 kg)	Minced beef	
6 lb (2·5 kg)	Minced onions	
1 lb (500 g)	Dripping	
	Seasoning	

1. Fry off the minced beef and onions in the dripping. Season with pepper and salt, stir frequently to avoid burning. Remove excess fat as it appears.

8 pints (4·5 litres)	Brown stock
	Mixed herbs
1	Crushed clove garlic

2. Add the stock and garlic. Sprinkle with herbs to taste, adjust the seasoning if necessary. Bring to the boil. Reduce the heat and simmer gently until the meat is cooked. More fat will appear during cooking, this must be removed as it appears. It may be necessary to add more stock to replace that lost through evaporation.

4 oz (120 g)	Plain flour
4 oz (120 g)	Freeze 'n' flow starch
	Water to mix

3. Mix the starch and flour to a smooth paste with cold water. Add to the mince stirring frequently. Bring back to the boil, reduce the heat and simmer for a few minutes to cook the flour.

4. Check for colour, seasoning and consistency.

5. Weigh 2 lb (1 kg) of minced beef into each 9½ in sq shallow foil container.

6. Lid and seal.

7. Put into blast freezer for 90 minutes.

8. Store in holding fridge until required.

Lid Notes
Description: Minced Beef and Onions
No. of portions: 6
Oven setting: 5
Oven time: 40 minutes
Lid on
Pack weight: 1 kg

MOUSSAKA (Economy) (120 PORTIONS)

16 lb (7 kg)	Peeled potatoes	1. Slice and par-boil the potatoes. Drain and cool.
10 lb (5 kg)	Onions	2. Peel and mince the onions through largest holes.
25 lb (12 kg)	Minced beef Dripping	3. Fry the minced beef and the onion in a bratt pan. Do not overcook the beef or it will be tough.
20 pints (11 litres)	Stock Seasoning	4. Add the stock and season. Bring to the boil, reduce the heat and cook gently for approx. 10 minutes.
1¼ lb (500 g) 1¼ lb (500 g)	Margarine Flour	5. Melt the margarine, add the flour and cook out the roux without colouring.
20 pints (8 litres)	Milk	6. Slowly stir in the milk and make a white sauce. Stir continuously until the sauce has thickened.
4 lb (2 kg) 1 tablespoon	Grated cheese Mustard Crushed clove garlic Seasoning	7. Add the cheese, mustard and garlic. Season to taste. Allow the sauce to cool.
8 lb (3·5 kg)	Tomatoes	8. Wipe and slice the tomatoes.
		9. Weigh 2 lb (1 kg) mince into each 9½ in sq deep container.
		10. Pour over 1 pint (0·5 litre) cheese sauce.
2 oz (50 g)	Chopped parsley	11. Arrange 3 oz (225 g) tomatoes over. Sprinkle with chopped parsley.
1½ lb (1 kg)	Grated cheese	12. Arrange the sliced potatoes over the top. Sprinkle with 1½ oz (40 g) grated cheese.
		13. Lid and seal.
		14. Put into freezer for 90 minutes.
		15. Store under refrigeration until required.

Lid Notes
Description: Moussaka
No. of portions: 8
Oven setting: 6
Oven time: 50 minutes
Lid off
Pack weight 1·75 kg

NOISETTE OF LAMB CHASSEUR (200 PORTIONS)

400 small (50–55 g)	Noisettes of lamb
190 lb (55 kg)	Middle lamb
3 lb (1½ kg)	Butter
5 lb (2 kg)	Chopped onions
3 lb (1·5 kg)	Sliced mushrooms
2 lb (1 kg)	Sliced celery
1½ lb (700 g)	Flour
1 lb (450 g)	Tomato puree
	Mixed herbs
	Seasoning
1 litre	Red wine
5 gals. (24 litres)	Stock
2½ lb (1100 g)	Freeze 'n' flow starch
	Gravy browning

1. Melt the butter in a bratt pan. When it is foaming, sauté the noisettes. Brown evenly on both sides, reduce the heat and cook gently until they are just under-done. Use bones from best ends to make stock.

2. Remove the noisettes from the butter and drain them on wire racks.

3. Prepare the vegetables and sweat them in the bratt pan in the dripping from the noisettes.

4. Pour in the red wine and allow it to reduce. Stir in the flour and tomato puree, add the seasoning and the herbs. Gently cook the flour, stirring all the time.

5. Gradually add the stock, stirring all the time, bring to the boil, reduce the heat and simmer until the vegetables are just cooked. Skim off excess fat and scum from the surface.

6. Reconstitute the starch with cold water and stir into the sauce. Simmer gently until the sauce clears and the starch is cooked. Colour with gravy browning.

7. Check the sauce for colour, seasoning and consistency.

8. Pack 20 noisettes in to each 9½ in sq deep pack.

9. Add 2 pints (1 litre) sauce.

10. Lid, seal and freeze.

11. Store in cold room until required.

Lid Notes
Description: Noisette of Lamb Chasseur
No. of portions: 10
Oven setting: 6
Oven time: 45 minutes
Lid on
Pack weight: 2 kg

ROAST BEEF (90 PORTIONS)

25 lb (12 kg)   Thick flank
                Seasoning
1 lb (500 g)   Dripping

1. Season the beef joints and brush well with melted dripping. Put into roasting tins. Put into pre-heated oven at regulo 7 for 20 minutes, reduce the oven setting to regulo 4 and continue cooking slowly. Baste from time to time.

2. When the beef is cooked remove it from the roasting tins and put on cooling trays to set.

3. Put the roasting tins on the solid top stoves and set the meat juices, taking care not to burn them. Pour off the fat into a pot and use as dripping.

8 pints (4·5 litres)   Water

4. Add hot water to the meat juices and stir until they have all dissolved. Sieve into a suitably sized pan and bring to the boil.

4 oz (120 g)   Freeze 'n' flow starch
               Water to mix
               Seasoning

5. Mix the starch with cold water to make a creamy paste. Stir this carefully into the meat juices. Simmer for 5 minutes to cook out. Season as required and correct the colouring. Cool quickly.

6. Slice the beef and put 10 slices of 2 oz (55 g) into each $9\frac{1}{2}$ in sq shallow foil container.

7. Add 1 pint (0·5 litre) of gravy.

8. Lid and seal. Freeze for 90 minutes.

9. Store in holding fridge until required.

Lid Notes
Description: Roast Beef
No. of portions: 10
Oven setting: 6
Oven time: 35 minutes
Lid on
Pack weight: 1050 g

ROAST LEG OF LAMB (100 PORTIONS)

30 lb (14 kg)	Leg of lamb (on bone)	1. Bone, roll and tie the legs of lamb. Use bones for stock.
1 lb (500 g)	Dripping Seasoning	2. Brush the legs of lamb with melted dripping, season, put on to wire racks in roasting tins.
		3. Put into a pre-heated oven on regulo 6 for 20 minutes. Reduce the heat to regulo 4 and continue cooking gently until the meat is well done. Baste occasionally with the juices from the roasting tins.
		4. When cooked, put the lamb on the racks in a cool place to set.
		5. Put the roasting tins on to a solid top stove and carefully set the juices. Pour off the excess fat into a pot and use for dripping.
8 pints (4·5 litres)	Brown stock	6. Add the stock to the roasting tins. Bring to the boil until all the sediment is dissolved. Sieve into a saucepan. Remove scum and excess fat.
4 oz (120 g)	Freeze 'n' flow starch Water to mix	7. Mix the starch to a smooth paste with cold water and stir into the gravy. Bring to the boil, simmer gently for a few minutes. Cool quickly. Check seasoning etc.
		8. Slice the lamb and put 20 × 25 g slices into each 9½ in sq shallow foil container.
		9. Add 1 pint (0·5 litre) of gravy.
		10. Lid, seal and put into freezer for 90 minutes.
		11. Store in holding fridge until required.

Lid Notes
Description: Roast Leg of Lamb
No. of portions: 10
Oven setting: 6
Oven time: 35 minutes
Lid on
Pack weight: 1 kg

ROAST LOIN OF PORK (96 PORTIONS)

30 lb (14 kg) — Loin of pork (including bone) / Seasoning

1. Season the joints and place on to long trays.

2. Switch on convector ovens and set on regulo 6.

3. Put the pork into oven and reduce temperature to regulo 4.

4. Set the oven timer for 45 minutes. When timer rings turn the joints over and set timer for further 45 minutes.

5. Test the joints to make sure that they are cooked.

6. When the joints are cooked remove them from the roasting tins and place on cooling racks.

7. Set the juices on the roasting trays and pour off the excess fat. Swill the roasting tins and put the juice into a saucepan.

8. Bring the juices to the boil and make the quantity up to 1 gal. (4·5 litres) with water.

12 oz (350 g) — Waxy maize / Water / Seasoning

9. Mix the waxy maize to a smooth paste with cold water and add to the stock. Season to taste, allow the gravy to cook.

10. Check for colour, flavour and consistency.

11. Carve the loin of pork and put 20 slices of 40 g into each 9½ in sq foil pack.

12. Add 1 pint (0·5 litre) of gravy.

13. Lid and seal. Place in freezer for 90 minutes.

Lid Notes
Description: Roast Loin of Pork
No. of portions: 10
Oven setting: 6
Oven time: 35 minutes
Lid on
Pack weight: 1·3 kg

ROAST SADDLE OF LAMB (120 PORTIONS)

15	Saddles (cut from chine and end)
1 lb (500 g)	Dripping
	Seasoning

1. Trim the breast and kidney fat from the saddles. Score the skin with point of sharp knife along the chine bone.

2. Brush the saddles with melted dripping, season them and put into roasting tins.

3. Put into a pre-heated oven on regulo 6 for about 15 minutes. Reduce heat to regulo 4 and continue until meat is cooked. Baste occasionally with the juices from the roasting tins.

4. When cooked, put the saddles on wire racks in a cool place to set.

5. When the saddles are cool upturn them one at a time and remove the under-fillets whole. Turn them over and with a sharp knife cut carefully down each side of the chine and remove the meat by following the bone with the blade of the knife. Place the meat back in place on the bones. Put the under-fillets back underneath.

6. Wrap them in foil, ensuring that they are air tight.

7. Label them.

8. Freeze for 90 minutes. Put into plastic bags and seal.

9. Store in cold room until required.

Lid Notes
Description: Roast Saddle of Lamb
No. of portions: 6–8
Oven setting: 6
Oven time: 40 minutes
Pack weight: 1·2 kg

ROAST SIRLOIN OF BEEF (50 PORTIONS)

25 lb (12 kg)   Sirloin on bone

1. Remove the fillet and bones from the strip loins. Roll and tie securely with string. Break bones and use for stock.

1 lb (500 g)   Dripping
               Seasoning

2. Season the beef and brush with melted dripping. Place on a rack in a suitably sized roasting tin.

Bones ⎫
Carrots ⎬ Stock
Onions ⎭

3. Put in a pre-heated oven at regulo 6. After 20 minutes, reduce the heat to regulo 4 and continue cooking until the beef is medium. Baste occasionally with the juices from the roasting tin.

4. When the beef is cooked put it on the rack in a cool place to set. When cold, slice.

5. Put the roasting tin(s) on a solid top stove and set the juices, taking care not to burn them. Pour off the fat into a pot and use as dripping.

1½ oz (50 g)   Flour

6. Stir the flour into the meat juices.

4½ pints (2·5 litres)   Stock

7. Gradually stir in the water and bring to the boil. Sieve into a suitably sized pan and bring to boil.

1½ oz (50 g)   Freeze 'n' flow starch
               Water to mix
               Seasoning

8. Mix the starch with cold water to a smooth paste. Stir into the gravy. Check seasoning, consistency (thin) and colour. Cool quickly.

9. Put 10 slices (70 g) into each 9½ in sq shallow foil container, add 1 pint (0·5 litre) gravy.

10. Lid, seal and freeze for 90 minutes.

Lid Notes
Description: Roast Sirloin of Beef
No. of portions: 10
Oven setting: 6
Oven time: 35 minutes
Lid on
Pack weight: 1·2 kg

SAUSAGE LYONNAISE (400 PORTIONS)

100 lb (45 kg)	Thick sausage
4 lb (2 kg)	Dripping
20 lb (10 kg)	Sliced onions
3 lb (1·5 kg)	Dripping
10 gals. (45 litres)	Water
2 lb (1 kg)	Tomato puree
	Bouquet garni
	Pepper and salt
4 lb (2 kg)	Freeze 'n' flow starch
3 lb (1·5 kg)	Flour
	Gravy browning

1. Arrange the sausages on to long trays and brush them with dripping. Brown them all over under the grill.

2. Peel and slice the onions. Lightly brown the onions in dripping in the bratt pan.

3. Add the sausages, 10 gals. (45 litres) water and the tomato puree and bouquet garni. Season, bring slowly to the boil, reduce heat and simmer gently for a few minutes to finish cooking the sausages.

4. Remove the sausages from the pan and allow them to cool.

5. Remove the surplus fat from the cooking liquor and adjust to 10 gals. (45 litres).

6. Mix the flour and freeze 'n' flow together and reconstitute with cold water. Stir it into the bratt pan. Bring slowly to the boil, reduce the heat and simmer gently until the flour is cooked out.

7. Check the sauce for colour, seasoning and consistency. Allow it to cool.

8. Pack 16 sausages into each $9\frac{1}{2}$ in sq shallow container.

9. Pour over 2 pints (1 litre) sauce.

10. Lid and seal.

11. Put into freezer for 90 minutes.

12. Store under refrigeration until required.

Lid Notes
Description: Sausage Lyonnaise
No. of portions: 8
Oven setting: 6
Oven time: 45 minutes
Lid on
Pack weight: 1·75 kg

SAUTÉ'D BEEF AND CARROTS (250 PORTIONS)

60 lb (28 kg)	Cubed beef	1. Season the meat and seal it quickly in hot dripping in a bratt pan.
3 lb (1 kg)	Dripping	
	Seasoning	
30 lb (14 kg)	Sliced carrots	2. Add the sliced vegetables and bouquet garni and fry until golden brown. Sprinkle with flour and cook the flour out.
12 lb (6 kg)	Sliced onions	
9 lb (4 kg)	Sliced celery	
1 oz	Monosodium glutamate	
60 pints (35 litres)	Brown stock	3. Add the stock, tomato puree and monosodium glutamate. Bring to the boil, reduce the heat to a simmer, skim off excess fat and any scum that appears. Simmer until meat is cooked.
1½ lb (700 g)	Tomato puree	
12 oz (350 g)	Freeze 'n' flow starch	4. Mix the starch, flour and a little cold water to a smooth paste and stir into the cooking liquor. Simmer gently until the flour is cooked.
12 oz (350 g)	Plain flour	
	Water to mix	

5. Check for seasoning, colour and consistency.

6. Remove from the bratt pan to packing station.

7. Weigh 2½ lb (1·75 kg) of sauté'd beef into each 9½ in sq shallow foil container.

8. Lid, seal and put into freezing tunnel for 90 minutes.

9. Remove the packs from the tunnel and store in holding fridges until required.

Lid Notes
Description: Sauté'd Beef and Carrots
No. of portions: 8
Oven setting: 6
Oven time: 45 minutes
Lid on
Pack weight: 1·75 kg

SAUTÉ'D BEEF AND MUSHROOMS (250 PORTIONS)

60 lb (28 kg)	Cubed beef
3 lb (1·5 kg)	Dripping
	Seasoning
24 lb (12 kg)	Sliced carrots
18 lb (8 kg)	Sliced onions
18 lb (8 kg)	Sliced mushrooms
9 lb (4 kg)	Celery
60 pints (35 litres)	Brown stock
1½ lb (700 g)	Tomato puree
	Mixed herbs
1 oz (25 g)	Monosodium glutamate
12 oz (350 g)	Freeze 'n' flow starch
12 oz (350 g)	Plain flour
	Water to mix

1. Season the meat and seal it quickly in hot dripping in a bratt pan.

2. Add the sliced vegetables and fry until golden brown. Add more dripping if required, sprinkle with a little flour to absorb excess fat.

3. Add the brown stock, tomato puree and monosodium glutamate. Sprinkle with herbs to taste. Adjust the seasoning. Bring to the boil, stirring occasionally to avoid burning. Remove any scum and fat as it rises to the surface. Reduce the heat and simmer until the meat is cooked. Stir occasionally.

4. Mix the starch and plain flour with cold water to form a smooth creamy paste. Stir into the bratt pan. Bring back to the boil, reduce the heat and simmer gently for a few minutes to cook the flour.

5. Check the seasoning, colour and consistency.

6. Remove from the bratt pan to the packing station.

7. Weigh 2½ lb (1·75 kg) of sauté'd beef into each 9½ in sq shallow foil container.

8. Lid, seal and put into freezing tunnel for 90 minutes.

9. Remove from the tunnel and store in holding fridge until required.

Lid Notes
Description: Sauté'd Beef and Mushrooms
No. of portions: 8
Oven setting: 6
Oven time: 45–50 minutes
Lid on
Pack weight: 1·75 kg

STEAK AND KIDNEY PIE (520 PORTIONS)

60 lb (28 kg)	Cubed steak	1. Put the cubed beef, kidney and onions into a bratt pan. Cover with brown stock or water (approx. 24 litres). Season and sprinkle with mixed herbs. Bring to the boil, reduce the heat and simmer gently until the meat is cooked. Skim off any fat and scum as it rises to the surface. Add more stock or water if and when necessary.
15 lb (7 kg)	Ox kidney	
10 lb (5 kg)	Sliced onions	
40 pints (24 litres)	Brown stock	
	Seasoning	
	Mixed herbs	

1 lb 12 oz (700 g)	Freeze 'n' flow starch	2. Mix the starch and flour to a smooth creamy paste with cold water. Add slowly to the bratt pan stirring continuously. Bring back to the boil, reduce the heat and simmer gently for about 5 minutes to cook the flour.
1 lb 12 oz (700 g)	Plain flour	
	Water to mix	

1 oz (25 g)   Monosodium glutamate   3. Add the monosodium glutamate; check for colour, seasoning and consistency.

60 lb (28 kg)   Shortcrust pastry   4. See method for preparing the shortcrust pastry.

5. Line the 9 in diameter pie foils with ½ lb (250 g) pastry.

6. Add 1 lb 2 oz (550 g) steak and kidney.

½ pint (250 cc)   Egg wash   7. Cover with 6 oz (200 g) pastry and brush with egg wash.

8. Cover with suitably labelled foil.

9. Freeze for 90 minutes.

10. Store in holding fridge until required.

Lid Notes
Description: Steak and Kidney Pie
Oven setting: 4
Oven time: 35 minutes
Lid off
Pack weight: 1 kg

STEAK AND KIDNEY PUDDING (480 PORTIONS)

60 lb (28 kg)	Cubed beef
15 lb (7 kg)	Ox kidney
10 lb (5 kg)	Onions (chopped)
	Seasoning
	Herbs
40 pints (24 litres)	Brown stock

1. Put the cubed beef, kidney and onions into a bratt pan. Cover with stock or water (approx. 5 gals.), season and sprinkle with mixed herbs. Bring to the boil, reduce the heat and simmer gently until the meat is cooked. Skim off any fat and scum as it rises to the surface. Add more stock or water if and when necessary.

1 lb 12 oz (700 g)	Freeze 'n' flow starch
1 lb 12 oz (700 g)	Plain flour
	Water to mix

2. Mix the starch and flour to a smooth creamy paste with cold water. Add slowly to the bratt pan stirring continuously. Bring back to the boil, reduce the heat and simmer gently for about 5 minutes to cook the flour.

1 oz (25 g)	Monosodium glutamate

3. Add the monosodium glutamate, check for colour, seasoning and consistency.

40 lb (20 kg)	Plain flour
2½ lb (1 kg)	Baking powder
20 lb (10 kg)	Suet
	Salt
13 pints (7½–8 litres)	Water to mix

4. Sieve the flour, baking powder and salt into large mixing bowl. Add the suet and blend the ingredients together. Use the dough hook.

5. Add sufficient cold water to make a soft dough (approx. 13 pints or 7½–8 litres). Do this as quickly as possible and avoid overmixing. Let the dough rest for 10 minutes.

6. Line the 9½ in round, vertical sided foil packs with 14 oz (400 g) pastry.

7. Add 2 lb (1 kg) steak and kidney.

8. Top with 7 oz (200 g) pastry.

9. Lid, seal and freeze.

Lid Notes
Description: Steak and Kidney Pudding
No. of portions: 8
Reheating: Steamer
Cooking time: 2½ hours
Lid on
Pack weight 1·6 kg (200 g portion)

STEAK AND MUSHROOM PIE (480 PORTIONS)

60 lb (28 kg)	Cubed steak
10 lb (4·5 kg)	Sliced onions
4 lb (2 kg)	Mushrooms
40 pints (24 litres)	Brown stock
	Seasoning
	Mixed herbs

1. Put the cubed beef, onions and mushrooms into a bratt pan. Cover with brown stock or water (approx. 40 pints or 24 litres). Season and sprinkle with mixed herbs. Bring to the boil, reduce the heat and simmer gently until the meat is cooked. Remove any scum and fat from the surface as it rises. Stir occasionally and add more stock or water as necessary during cooking.

1 lb 12 oz (700 g)	Freeze 'n' flow starch
1 lb 12 oz (700 g)	Plain flour
	Water to mix

2. Mix the starch and flour to a smooth creamy paste with cold water. Add slowly to the bratt pan, stirring continuously. Bring back to the boil, reduce the heat and simmer gently for about 5 minutes to cook the flour.

60 lb (28 kg)	Shortcrust pastry

3. See method for preparation of shortcrust pastry.

4. Line 9 in diameter pie foils with 8 oz (250 g) pastry.

5. Add 1 lb 2 oz (550 g) steak and mushrooms.

½ pint (250 cc)	Egg wash

6. Cover with 7 oz (200 g) pastry and brush with egg wash.

7. Cover with suitably labelled foil.

8. Freeze for 90 minutes.

9. Store in holding fridge until required.

Lid Notes
Description: Steak and Mushroom Pie
Oven setting: 4
Oven time: 35 minutes
Lid off
Pack weight: 1 kg

STEAK AND ONION PIE (450 PORTIONS)

60 lb (28 kg)	Cubed steak	1. Put the cubed beef and onions into a bratt pan. Cover with brown stock or water (40 pints or 24 litres). Season and sprinkle with mixed herbs. Bring to the boil, reduce the heat and simmer gently until the meat is cooked. Remove any fat and scum as it rises to the surface. Stir occasionally and add more stock or water as necessary during cooking.
12 lb (6 kg)	Sliced onions	
40 pints (24 litres)	Brown stock	
	Seasoning	
	Mixed herbs	
1 lb 12 oz (700 g)	Freeze 'n' flow starch	2. Mix the starch and flour to a smooth creamy paste with cold water. Add slowly to the bratt pan stirring continuously. Bring back to the boil and reduce heat. Simmer gently for about 5 minutes to cook the flour.
1 lb 12 oz (700 g)	Plain flour	
	Water to mix	
1 oz (25 g)	Monosodium glutamate	3. Add the monosodium glutamate, check for colour, seasoning and consistency.
60 lb (28 kg)	Shortcrust pastry	4. See method for preparation of shortcrust pastry.
		5. Line 9 in diameter pie foils with 8 oz (250 g) pastry.
		6. Add 1 lb 2 oz (550 g) steak and onion.
$\frac{1}{2}$ pint (250 cc)	Egg wash	7. Top with 7 oz (200 g) pastry and brush with egg wash.
		8. Cover with suitably labelled foil.
		9. Freeze for 90 minutes.
		10. Store in holding fridge until required.

Lid Notes
Description: Steak and Onion Pie
Oven setting: 4
Oven time: 35 minutes
Lid off
Total weight: 1 kg

STUFFED SHOULDER OF LAMB (80 PORTIONS)

22 lb (10 kg)	Shoulder of lamb	1. Carefully remove all the bones from the shoulders. Use them to make a stock.
5 lb (2 kg)	Stuffing	2. Stuff and roll the shoulders. Tie them securely and put on to racks into roasting tins.
1 lb (500 g)	Dripping Seasoning	3. Brush the shoulders with melted dripping, season them and put them in a pre-heated oven on regulo 6 for 15 minutes. Reduce to regulo 4 and continue until the meat is cooked. Baste occasionally with the juices from the roasting tins.
		4. When cooked put the shoulders on the racks in a cool place to set.
		5. Put the roasting tins on a solid top stove and carefully set the juices. Pour off the fat into a pot and use as dripping.
8 pints (4·5 litres)	Brown stock	6. Pour the stock into the roasting tins, stir until all the sediment is dissolved. Sieve into a saucepan.
4 oz (120 g) 4 oz (120 g)	Freeze 'n' flow starch Plain flour Cold water to mix Seasoning	7. Mix the starch and flour to a smooth paste with cold water. Gradually pour this into the stock, stirring all the time. Bring to the boil and simmer gently for a few minutes to cook the flour. Check for colour, seasoning and consistency. Cool quickly.
		8. Remove the string from the shoulders and slice.
		9. Put 20 slices of 1 oz (25 g) each into each 9½ in sq shallow foil container.
		10. Add 1 pint (0·5 litre) gravy.
		11. Lid and seal.
		12. Freeze for 90 minutes.
		13. Store in holding fridge until required.

Lid Notes
Description: Stuffed Shoulder of Lamb
No. of portions: 10
Oven setting: 6
Oven time: 35 minutes
Lid on
Pack weight: 1 kg

STUFFING FOR PORK (330 PORTIONS)

5 lb (2·5 kg)	Onions	1. Peel and mince the onions, using the largest mincer plate.
		2. Cover with water and boil.
10 lb (5 kg)	Breadcrumbs	3. Put the dry ingredients into a mixing bowl and blend at a slow speed.
3 oz (100 g)	Sage	
5 oz (150 g)	Freshly chopped parsley	
	Seasoning	
		4. Add the boiled onions and their cooking liquor.
2 lb (1 kg)	Pork dripping	5. Add the pork dripping.
	Stock to mix	6. Add stock to achieve correct consistency.
		7. Check for seasoning.
		8. Using the smallest ice cream scoop, portion the stuffing into 1 oz (25 g) balls.
		9. Put 24 portions into each $9\frac{1}{2}$ in sq shallow container.
		10. Lid and seal.
		11. Put into blast-freezer for 90 minutes.
		12. Put into cold storage until required.

Lid Notes
Description: Stuffing Balls
No. of portions: 24
Oven setting: 5
Oven time: 35 minutes
Lid on
Pack weight: 600 g

Chicken, Duck and Game

CHICKEN À LA KING (400 PORTIONS)

20 × 5 lb (2·5 kg)	Boiling fowl	1. Defrost the boiling fowl and remove the giblets.
		2. Just cover the boiling fowl with salted water in a bratt pan. Bring to the boil and simmer gently until cooked.
6 lb (3 kg) 6 lb (3 kg) 6 lb (3 kg) 6 lb (3 kg)	Sliced onions Capsicums Mushrooms Red peppers	3. Meanwhile clean and slice the onions, mushrooms, red and green peppers.
		4. Remove the boiling fowl from the bratt pan, place on to wire racks and cool a little.
		5. Drain the stock from the bratt pan into a suitably sized saucepan, skim off the excess fat.
		6. Using some of the fat from the stock sweat off the sliced vegetables in the bratt pan. Take care not to colour the vegetables.
$2\frac{3}{4}$ lb (1 kg)	Flour	7. Stir in the flour and allow to cook without colouring. This will absorb the fat.
9 gals. (40 litres)	Chicken stock	8. Stir in the chicken stock; the amount of stock obtained may have to be made up with water. Bring to the boil, reduce heat to simmer.
	Pepper Salt Mixed herbs	9. Add the herbs and seasoning to the sauce. Stir the sauce occasionally and remove the scum and fat as it appears. Allow the sauce to cook gently.
1 litre	White wine	10. Reduce the wine by a half and add to the sauce.
6 lb (2·75 kg)	Milk powder	11. Reconstitute the milk powder and add to the sauce.
4 lb (2 kg)	Freeze 'n' flow starch	12. Reconstitute the freeze 'n' flow and stir into the sauce, stirring all the time until the starch is cooked. Check the sauce for colour, flavour and consistency.
		13. Cut the chicken into cubes, after removing the flesh from the carcasses, while the sauce is cooling.
		14. Mix the chicken with the sauce. Again check for flavour and consistency.
		15. Weigh 4 lb (1·75 kg) of chicken à la king into each $9\frac{1}{2}$ in sq deep foil container.

Lid Notes
Description: Chicken à la King
No. of portions: 8
Oven setting: 6
Oven time: 45 minutes
Lid on
Pack weight: 1·75 kg

16. Seal with suitably labelled lids.

17. Put into freezer tunnel for 90 minutes.

18. Put into cold store until required.

85

CHICKEN AND HAM PIE (600 PORTIONS)

12 × 5½ lb (2·5 kg)	Boiling fowl	1. Defrost the fowl and remove the giblets.
		2. Put the fowl into a bratt pan and just cover with salted water. Bring to the boil, reduce the heat and simmer gently until cooked. Remove the fowl from the bratt pan and cool on wire racks.
15 lb (7 kg)	Diced ham	3. Boil the ham in unsalted water, drain and allow to cool.
5 gals (24 litres) 6 lb (2·5 kg) 6 lb (2·5 kg) 6 lb (2·5 kg)	Chicken stock Onions Carrots Mixed herbs Green peppers	4. Clean and slice the onions and carrots and peppers and cook them lightly in the chicken stock. Add the mixed herbs to taste.
3 lb (1·25 kg) 3 × A2	Frozen peas Pimentoes	5. Add the peas and the sliced pimentoes to the bratt pan.
3 lb (1·25 kg)	Milk powder	6. Reconstitute the milk powder and stir into the bratt pan.
1½ lb (750 g) 2¼ lb (1 kg)	Flour Freeze 'n' flow starch	7. Mix the flour and freeze 'n' flow together and reconstitute with water and stir into the bratt pan, bring slowly to the boil, stirring all the time. Reduce the heat and allow the sauce to simmer until the flour is cooked.
	Pepper and salt	8. Check the sauce for colour, flavour and consistency and allow to cool.
		9. Meanwhile remove the flesh from the chicken and cut it into regular dice.
		10. Blend the diced ham and chicken into the sauce. Check colour, flavour and consistency.
60 lb (28 kg)	Shortcrust pastry	11. Line the 9 in diameter pie foils with 8 oz (250 g) pastry.
		12. Add 1 lb 2 oz (550 g) chicken filling.
		13. Top with 7 oz (200 g) pastry.
½ pint (250 cc)	Egg wash	14. Brush with egg wash.
		15. Cover with suitably labelled foil.
		16. Put into freezer for 90 minutes.
		17. Store under refrigeration until required.

Lid Notes
Description: Chicken and Ham Pie
No. of portions: 6
Oven setting: 4
Oven time: 35 minutes
Foil cover off
Pack weight: 1 kg

CHICKEN CHASSEUR (400 PORTIONS)

100 × 2 lb (1 kg)	Chickens	1. Defrost the chickens and remove the giblets.
		2. Roast the chickens carefully in the convector ovens. Put the giblets into a saucepan, bring to the boil and simmer gently.
10 lb (5 kg) 5 lb (2·5 kg) 6 lb (3 kg)	Onions Mushrooms Celery	3. Meanwhile prepare the onions, mushrooms and celery and slice them evenly.
		4. Put the chickens to cool and de-glace the roasting tins.
3 lb (1·5 kg)	Dripping	5. Take 3 lb (1·5 kg) of the dripping from the roasting tins and put it in a bratt pan. Sweat the vegetables without colouring them.
2 lb (1 kg) 2 lb (1 kg) 2 oz (50 g) 4 oz (100 g) 1 litre 10 gals. (45 litres)	Flour Tomato puree Mixed herbs Chopped parsley Seasoning Red wine Chicken stock (including giblet stock and sediment from roasting tins)	6. Sprinkle the flour into the bratt pan and stir. Cook the flour slowly then add the tomato puree, herbs, seasoning and wine. Gradually stir in the stock. Bring to the boil, stirring all the time, reduce the heat and simmer gently until the vegetables are just cooked. Remove scum as it appears.
		7. While the sauce is cooking, portion the chickens into 4 portions each. Put 8 portions into each 9½ in sq deep foil container.
3½ lb (1·5 kg)	Freeze 'n' flow starch	8. Reconstitute the freeze 'n' flow and stir into the sauce, bring to the boil, stirring all the time, reduce the heat to a simmer and cook the sauce.
		9. Check the sauce for colour, flavour and consistency. Allow it to cool.
		10. Pour 1½ pints (1 litre) sauce on to each pack of chicken portions.
		11. Lid and seal and freeze.
		12. Put into cold storage until required.

Lid Notes
Description: Chicken Chasseur
No. of portions: 8
Oven setting: 6
Oven time: 50 minutes
Lid on
Pack weight: 2 kg

CHICKEN CROQUETTES (350 PORTIONS)

15 × 5 lb (2·5 kg)	Boiling fowl
5 lb (2·5 kg)	Minced onions
8 lb (3·5 kg)	Breadcrumbs Mixed herbs Seasoning
12 pints (6·5 litres)	Chicken stock
8 oz (250 g) 8 oz (250 g)	Waxy maize Plain flour
	Seasoned flour Egg wash Breadcrumbs

1. Boil the fowl and save the stock. Allow the fowl to cool. Strip the meat from the bones and put through the mincer.

2. Boil the minced onions with a little seasoned water.

3. Blend the minced chicken, onions, breadcrumbs and mixed herbs in a mixing bowl. Season lightly.

4. Bring the chicken stock to the boil.

5. Mix the flour and waxy maize to a smooth paste with cold water. Add to the chicken stock, bring back to the boil, stirring thoroughly. Reduce heat and simmer gently until the flour has cooked.

6. Add the sauce gradually to the chicken and mix thoroughly on a slow speed. Check the seasoning.

7. With a size 24 scoop, portion the chicken and pass through seasoned flour, egg wash and breadcrumbs.

8. Re-shape the croquettes if necessary and pack 16 croquettes to a $9\frac{1}{2}$ in sq foil pack (2 croquettes per portion).

9. Lid and seal.

10. Put into freezer for 90 minutes.

Lid Notes
Description: Chicken Croquettes
No. of portions: 8
Deep fry at 350 °F for 5–7 minutes.

CHICKEN FRICASSÉ (400 PORTIONS)

20 × 5 lb (2·5 kg)	Boiling fowl	1. Defrost the fowl and remove the giblets.
		2. Put the fowl into a bratt pan, cover with salted water and bring to the boil. Reduce the heat and simmer until cooked. Put the giblets into a saucepan, cover with water and make a stock.
15 lb (7 kg)	Diced bacon	3. Boil the bacon and drain.
8 lb (3·5 kg)	Onions	4. Clean and slice all the vegetables.
3 lb (1·5 kg)	Mushrooms	
8 lb (3·5 kg)	Carrots	
2 heads (1 kg)	Celery	
		5. When the fowl are cooked cool them on wire racks. Save some of the chicken fat for the next stage.
3 lb (1·25 kg)	Chicken fat	6. Sweat the vegetables in chicken fat in the bratt pan.
$2\frac{3}{4}$ lb (1 kg)	Flour	7. Sprinkle in the flour to absorb the fat and to form a roux. Cook the roux without colouring.
9 gals. (40 litres)	Chicken stock	8. Add the giblet stock to the stock from the fowl to make up 9 gals. (40 litres). Stir this gradually into the roux. Bring to the boil, reduce the heat and simmer gently until the vegetables are just cooked.
	Pepper	9. Season the sauce, add the mixed herbs and chopped parsley. Reconstitute the milk powder and add to the sauce.
	Salt	
	Mixed herbs	
2 oz (50 g)	Chopped parsley	
6 lb (2·75 kg)	Milk powder	
4 lb (2 kg)	Freeze 'n' flow starch	10. Reconstitute the freeze 'n' flow and stir into the sauce. Bring to the boil, stirring all the time. Reduce the heat and simmer until the starch is cooked. Check the sauce for colour, seasoning and consistency. Allow the sauce to cool.
		11. While the sauce is cooling remove the chicken flesh from the carcasses and cut into regular $\frac{3}{4}$ in dice. Add the diced chicken and ham to the sauce.
		12. Weigh 4 lb (1·75 kg) of fricassé into each $9\frac{1}{2}$ in sq shallow container.
		13. Lid and seal.
		14. Put into freezer for 90 minutes.
		15. Store under refrigeration until required.

Lid Notes
Description: Chicken Fricassé
No. of portions: 8
Oven setting: 6
Oven time: 45 minutes
Lid on
Pack weight: 1·75 kg

COQ AU VIN (400 PORTIONS)

25 × 2 lb (1 kg)	Chickens

1. Roast chickens. Remove from roasting tins and cool.
2. Boil the giblets, simmer gently to stop clouding.
3. Set the sediment in the roasting tins and pour off the fat. De-glace with the chicken stock and strain back into a saucepan. Toss in a few parsley stalks, celery tops and a bay leaf.

6 lb (3 kg)	Cubed bacon

4. Sweat the bacon in a bratt pan in the chicken fat.

4 lb (2 kg)	Chopped onions
2 lb (1 kg)	Sliced mushrooms

5. Toss the onions and mushrooms into the bacon and sweat. Pour off excess fat.

2 bottles (1 litre)	Red wine
1 gill (100 cc)	Brandy
10 gals (45 litres)	Chicken stock

6. Add the wine, brandy and chicken stock, bring to the boil, reduce the heat and simmer gently until the bacon is cooked.
7. Drain the cooking liquor into a suitably sized saucepan and bring to the boil.

2¼ lb (1 kg)	Flour
4 lb (2 kg)	Freeze 'n' flow starch
	Water to mix

8. Mix the flour and starch to a smooth thin paste with cold water and stir into the saucepan. Bring back to the boil, stirring continuously. Reduce heat and simmer gently for 5 minutes to cook the starch.

	Seasoning
	Gravy browning

9. Check the sauce for colour, flavour and consistency, correct as necessary.
10. Portion the chickens.
11. Put 8 portions of chicken into each 9½ in sq deep foil container.
12. Sprinkle with the bacon, onions and mushrooms.
13. Add 2 × 15 oz (1 litre) ladles of sauce.
14. Lid and seal.
15. Freeze for 90 minutes.
16. Store in cold room until required.

Lid Notes
Description: Coq au Vin
No. of portions: 8
Oven setting: 6
Oven time: 45 minutes
Lid on
Pack weight: 1·8 kg

CURRIED CHICKEN (400 PORTIONS)

100 × 2 lb (1 kg)	Chickens	1. Defrost the chickens and remove the giblets.

2. Roast the chickens carefully in the convector ovens. Put the giblets into a saucepan, bring to the boil, skim, simmer gently.

16 lb (7 kg)	Onions

3. Meanwhile, prepare and slice the onions.

4. Put the chickens to cool and de-glace the roasting tins.

3 lb (1·5 kg)	Dripping
5 lb (2 kg)	Curry powder

5. Lightly fry off the onions in chicken fat taken from the roasting tins. Add the curry powder and cook.

2 lb (1 kg)	Flour
2 lb (1 kg)	Tomato puree

6. Sprinkle in the flour, to take up the excess fat and cook. Stir in the tomato puree.

10 gals. (45 litres)	Giblet stock

7. Make up 10 gals. (45 litres) of chicken stock with the residue from the roasting tins and the giblet stock and add gradually to the roux.

8. Bring the sauce to the boil, remove scum and excess fat as it appears. Reduce the heat and simmer gently.

3 lb (1·25 kg)	Desiccated coconut
4 lb (2 kg)	Mango chutney
2 lb (1 kg)	Marmalade
2 × A10	Apples
	Pepper and salt

9. Slice the apples and add to the sauce together with the marmalade, chutney and coconut. Season to taste.

3½ lb (1·5 kg)	Freeze 'n' flow starch
6 lb (3 kg)	Sultanas

10. Reconstitute the freeze 'n' flow and stir into the sauce, stirring all the time bring the sauce to the boil. Reduce heat and simmer gently until the starch is cooked. Add the sultanas. Again test the sauce for colour, flavour and consistency; correct if necessary. Allow the sauce to cool.

11. Chop each chicken into 4 portions and put 8 portions into each $9\frac{1}{2}$ in sq deep pack.

12. Pour $1\frac{1}{2}$ pints (1 litre) of curry sauce into each pack.

13. Lid and seal.

14. Freeze for 90 minutes.

15. Store under refrigeration until required.

Lid Notes
Description: Curried Chicken
No. of portions: 8
Oven setting: 6
Oven time: 50 minutes
Lid on
Pack weight: 2 kg

DUCKLING À L'ORANGE (100 PORTIONS)

20 × 4½ lb (2 kg)   Duckling

1. Defrost the ducks and remove the giblets.

2. Pre-heat the ovens on regulo 6. Place the ducks on to roasting trays and brush lightly with lard. Put into ovens to cook. Reduce the temperature to regulo 4 after 30 minutes; continue at this temperature until the ducks are cooked.

3. Put the giblets into a saucepan, cover with water and bring to the boil.

4. Put the cooked ducks on to wire racks to cool.

1½ lb (700 g)	Duck fat
1½ lb (700 g)	Carrots
1½ lb (700 g)	Onions
½ lb (250 g)	Celery
½ lb (250 g)	Tomato puree
	Bouquet garni
1½ lb (700 g)	Flour

4 gals. (18 litres)   Stock

5. Pour 1½ lb (1 kg) of the fat from the roasting tins into a saucepan. Fry the sliced carrots, onions and celery. Add the flour and tomato puree and cook carefully without burning.

6. Swill out the roasting tins and measure the amount of stock obtained. Make this quantity up to 4 gals. (18 litres) with water. Stir this gradually into the roux. Bring to the boil, reduce the heat and simmer until the flour is cooked. Remove excess fat and scum as it appears.

20	Oranges
6	Lemons
	Pepper and salt

7. Thinly peel the oranges and cut the peel into ⅛ in strips. Squeeze the juice of the oranges into the sauce. Season the sauce to taste. Reduce the sauce by 3 pints (1·75 litres).

8. Strain the sauce into a clean saucepan.

1½ lb (700 g)   Freeze 'n' flow starch

9. Reconstitute the freeze 'n' flow with cold water and stir into the sauce. Bring slowly to the boil, reduce the heat and simmer until the starch is cooked.

½ pint (250 cc)   Brandy
                   Gravy browning

10. Add the brandy and check the sauce for colour, flavour and consistency. Correct if necessary.

11. Allow the sauce to cool.

12. Cut the ducks into 5 portions each and arrange 5 portions into each 9½ in sq deep container.

13. Sprinkle with the strips of orange peel.

14. Pour over 1½ pints (1 litre) sauce.

15. Lid and seal.

16. Put into freezer for 90 minutes.

17. Store in holding fridge until required.

Lid Notes
Description: Duckling à l'orange
No. of portions: 5
Oven setting: 6
Oven time: 45 minutes
Lid on
Pack weight: 1·8 kg

LIVER PATE (220 × 50 g PORTIONS)

7 lb (3 kg)	Pork dripping
4 lb (2 kg)	Chicken livers
4 lb (2 kg)	Onions
3	Cloves garlic
	Mixed herbs
	Seasoning
½ pint (250 cc)	Brandy
3 tins	Chopped truffles

1. Chop the onions and crush the garlic. Sauté with the chicken livers in half the pork dripping. Sprinkle with mixed herbs. Make sure that the livers are cooked. Allow them to cool.

2. Pass them through the fine mincer plate at least 4 times to obtain a smooth paste.

3. Stir in the rest of the dripping, the brandy and truffles (optional) and season to taste.

4. Weigh 2 oz (50 g) into a suitable smooth-sided container with lid.

5. Lid and seal and label.

6. Put into freezer for 90 minutes.

7. Store under refrigeration until required.

Lid Notes
Description: Liver Pate
Defrost and serve

ROAST CHICKEN (400 PORTIONS)

100 × 2 lb (1 kg)	Chickens	1. Defrost the chickens and remove the giblets.
	Dripping	2. Put the chickens into roasting tins and brush them with melted dripping.
		3. Pre-heat the convector ovens on regulo 7.
		4. Put the chickens into the ovens. After 10 minutes reduce heat to regulo 5 setting. Continue cooking at this temperature until the chickens are properly cooked.
6 gals. (27 litres)	Water	5. Put the giblets into a saucepan, cover with water and bring to the boil, reduce the heat and simmer gently. Remove the scum as it appears.
		6. Put the chickens to cool on racks.
		7. Pour the excess fat from the roasting tins, then, using the giblet stock, de-glace the tins. The resultant juice will make the gravy.
		8. Strain the juice from the roasting tins into a $7\frac{1}{2}$ gal. saucepan and bring to the boil.
1 lb (450 g) $1\frac{1}{2}$ lb (1 kg)	Flour Freeze 'n' flow starch	9. Mix the flour and freeze 'n' flow together and reconstitute with cold water. Stir this into the chicken stock, adjust the quantity to $6\frac{1}{2}$ gals. (30 litres) with water if necessary. Bring the gravy to the boil stirring all the time. Reduce the heat and simmer gently until the flour is cooked.
	Pepper and salt Gravy browning	10. Season the gravy and colour with gravy browning. Check for colour, flavour and consistency. The gravy should be thin. Allow the gravy to cool.
		11. Portion the chickens, 4 portions to a bird, and arrange 8 portions into each $9\frac{1}{2}$ in sq deep container.
		12. Pour over 1 pint ($\frac{1}{2}$ litre) gravy.
		13. Lid and seal.
		14. Put into freezer for 90 minutes.
		15. Store under refrigeration until required.

Lid Notes
Description: Roast Chicken
No. of portions: 8
Oven setting: 6
Oven time: 45 minutes
Lid on
Pack weight: 1·5 kg

ROAST GUINEA FOWL (100 PORTIONS)

50	Guinea fowl	1. Remove the giblets from the birds and wrap in larding fat if necessary.
		2. Put the birds into roasting tins and cook in a moderate oven until they are properly cooked. Baste them occasionally.
1 gal. (4·5 litres)	Water	3. Put the giblets into a saucepan and cover with water. Bring to the boil reduce the heat and simmer, removing scum from the surface of the stock as it appears.
		4. When the Guinea fowl are cooked, remove them from the roasting tins and put to cool.
		5. Put the roasting tins on top of a hot stove, set the residue and separate it from the excess fat. Do not burn.
		6. Pour the excess fat off the roasting tins.
		7. Strain the giblet stock into the residue and bring to the boil, stirring all the time to dissolve the meat extracts.
	Seasoning	8. Strain the gravy from the roasting tins into a saucepan and season to taste. Colour if necessary.
		9. Cool the gravy.
		10. Cut the Guinea Fowl in half by inserting a chopping knife through the rear cavity and cutting through the back bone. Turn the bird over and cut through the breast bone. Remove the rib cage and the back bone.
		11. Put 4 portions into a $9\frac{1}{2}$ in sq foil pack.
		12. Add 10 oz (250 cc) of gravy to each container.
		13. Lid and seal.
		14. Put into freezing tunnel for 90 minutes.
		15. Store in holding fridge until required.

Lid Notes
Description: Roast Guinea Fowl
No. of portions: 4
Oven setting: 5
Oven time: 40 minutes
Lid on
Pack weight: 1·5 kg

ROAST PHEASANT (100 PORTIONS)

50	Pheasants (ready dressed)	1. If the pheasants have not been wrapped in larding fat when purchased, lard them.
		2. Place in roasting tins and cook in a moderate oven regulo 6 until they are properly cooked. Baste them occasionally.
1 gal. (4·5 litres)	Water	3. Put the giblets in a saucepan, cover with water and bring to the boil. Reduce the heat and simmer, removing any scum from the surface of the stock as it appears.
		4. When cooked, remove the pheasants from the roasting tins and put to cool.
		5. Put the roasting tins on top of a hot stove to set the residue and separate it from the fat. Do not burn.
		6. Pour the excess fat off the roasting tins.
		7. Strain the giblet stock into the residue and bring to the boil, stirring all the time to dissolve the residue.
	Seasoning	8. Strain the gravy from the roasting tins into a saucepan and season to taste. Colour if necessary.
		9. Cool the gravy.
		10. Cut the pheasants into two by cutting through the backbone and the breast bone. Remove the backbone and rib cage.
		11. Put into foil packs 4 portions to a $9\frac{1}{2}$ in sq deep container.
		12. Add 10 oz (250 cc) of gravy to each container.
		13. Lid and seal.
		14. Put into freezing tunnel for 90 minutes.
		15. Store in holding fridge until required.

Lid Notes
Description: Roast Pheasant
No. of portions: 4
Oven setting: 5
Oven time: 40 minutes
Lid on
Pack weight: 1·5 kg

Fish

AMERICAN FISH PIE (104 PORTIONS)

1 stone (6·25 kg)	Cod or haddock	1. Cut the fish into regular sized pieces and arrange in buttered trays.
1 stone (6·25 kg)	Smoked haddock	
	Seasoning	2. Season the fish and cover with water. Cover with foil and poach gently in a moderate oven.
	Water	
1 gal. (4·5 litres)	Fish stock from the fish	3. Discard the stock off the smoked haddock and strain the fish stock from the cod into a saucepan; one gallon will be required.
1 gal. pack (450 g)	Milk powder	4. Bring the fish stock to the boil and stir in the milk powder.
6 oz (175 g)	Freeze 'n' flow	5. Mix the flour and the starch to a thin smooth paste with cold water. Stir into the fish stock. Bring back to the boil stirring all the time. Reduce the heat and simmer for 5 minutes to cook the starch.
6 oz (175 g)	Plain flour	
	Water to mix	
	Seasoning	6. Stir in the crushed clove of garlic and the parsley. Season to taste. Check the sauce for flavour and consistency, correct if necessary.
1 clove	Crushed garlic	
4 oz (100 g)	Fresh chopped parsley	
		7. Put the cooked fish into the mixing bowl and add the sauce gradually, mix together on slow speed. Add sufficient sauce to form a soft paste. Check for seasoning again.
3 lb (1·35 kg)	Potato powder	8. Mix the potato powder with the boiling water following manufacturers instructions. Season to taste and add the eggs. Beat well.
4 pints (2·25 litres)	Boiling water	
	Seasoning	
6	Eggs	
		9. Weigh 3 lb (1·35 kg) fish mixture into each 9½ in sq deep container.
		10. Pipe potato over the top (1 lb 8 oz or 675 g), leaving a gap down the centre.
3 lb (1·35 kg)	Sliced tomatoes	11. Arrange 4 oz (100 g) sliced tomatoes down the centre gap.
		12. Lid and freeze for at least 90 minutes.
		13. Put into cold storage until required.

Lid Notes
Description: American Fish Pie
No. of portions: 8
Oven setting: 5
Oven time: 50 minutes
Lid off
Pack weight: 2·125 kg

CRAWFISH AMERICAN STYLE (80 PORTIONS)

80 × 5–6 oz (140–170 g)	Crawfish tails	1. Remove the flesh from the tails in one piece. Cut the flesh into $\frac{3}{4}$ in slices.
1 lb (500 g) 3 lb (1·35 kg)	Butter Finely chopped onions	2. Melt the butter in a saucepan and sweat the finely chopped onions under a lid and without colouring.
		3. Add the crawfish tails and allow to cook gently for about 10 minutes.
3 gills (425 cc)	Brandy	4. Add the brandy and reduce it.
1 lb (500 g) 9 lb (4 kg)	Tomato puree Concassé tomatoes	5. Add the tomato puree and concassé tomatoes. Simmer until the tails are cooked.
6 pints (3·5 litres)	Dry white wine	6. Reduce the wine and add it to the sauce.
	Cayenne pepper Salt	7. Season the sauce to taste with cayenne pepper and salt.
		8. Weigh 2 lb (900 g) crawfish tails into each $9\frac{1}{2}$ in sq shallow container.
		9. Mask with one 15 fl. oz (425 cc) ladle of sauce.
2 oz (50 g)	Finely chopped parsley	10. Sprinkle with freshly chopped parsley.
		11. Lid and seal.
		12. Put into freezer for 90 minutes.
		13. Store in holding fridge until required.

Lid Notes
Description: Crawfish American Style
No. of portions: 8
Oven setting: 5
Oven time: 40 minutes
Lid on
Pack weight: 1·325 kg

FILLET OF FISH BONNE FEMME
(90–96 PORTIONS)

2 stones (12·5 kg)	Fish (cod, haddock, sole,* mock halibut)	1. Cut the fish into 4 oz (125 g) portions and arrange in a buttered dish.
2½ lb (1 kg)	Mushrooms	2. Wash, drain and slice the mushrooms.
½ lb (250 g)	Margarine	3. Place the mushrooms into a pan with the margarine and water. Season them. Put a lid on the pan and gently cook the mushrooms.
½ pint (250 cc)	Water	
	Seasoning	
2 lb (1 kg)	Finely chopped onions	4. Sprinkle the fish with finely chopped onions, seasoning and cover with fish stock or water and milk. Add a bayleaf and cook in a moderate oven. *Do not overcook.*
1 gal. (4·5 litres)	Water or fish stock	
1½ gals. (7·5 litres)	Milk	
	Bayleaf	
	Seasoning	
		5. Drain the cooking liquor from the fish into a saucepan, add the juice from the mushrooms. Adjust to 14 pints (8 litres) and bring to the boil.
1 pint (0·5 litre)	Medium dry white wine	6. Reduce the wine with a few peppercorns and lemon juice to one-third its volume. Add to the fish stock.
	Juice of one lemon	
	Few peppercorns	
12 oz (350 g)	Flour	7. Mix the flour and waxy maize to a smooth paste with cold water. Stir into the boiling fish liquor, bring back to the boil, and cook the flour gently, stirring all the time.
1 lb 4 oz (550 g)	Waxy maize	
	Water to mix	
1 pint (0·5 litre)	Cream	8. Place the pan on the side of the stove. Add the cream and butter, stir until all the butter is absorbed.
2 oz (50 g)	Butter	
		9. Check the sauce for colour, flavour and consistency.
		10. Put 8 portions of fish into each 9½ in sq shallow foil pack.
		11. Put 1 dessertspoonful of mushrooms on to each portion.
		12. Cover with 2 × 15 fl. oz (1 litre) ladles of sauce.
2 oz (50 g)	Chopped parsley	13. Sprinkle with chopped parsley.
		14. Lid and seal.
		15. Place in freezer.

* Sole should be filleted and skinned. Make fish stock from bones and skins. Use 3 oz portions of fish. Gives 140–150 portions.

Lid Notes
Description: Fillet of Fish Bonne Femme
No. of portions: 8
Oven setting: 6
Oven time: 40 minutes approx.
Lid off if served from pack
Lid on if served from flat
Pack weight: 2 kg

This is an alternative recipe to the one costed in Fig. 24.

FILLET OF FISH BRETONNE (96 PORTIONS)

2 stone (12·5 kg)	Cod or haddock (Fillets)	1. Cut the fish into 4 oz (100 g) portions and arrange in buttered trays.
2 lb (1 kg)	Sliced mushrooms	2. Blanch the mushrooms in salted boiling water and drain.
1½ gal. (7·5 litres) 1 gal. (4·5 litres)	Milk Fish stock Seasoning Bayleaves	3. Mix the milk and fish stock together and pour over the fish, just covering it. Sprinkle the mushrooms over the fish, add a bayleaf to each tray of fish and season.
		4. Cover with foil and poach gently in a moderate oven.
2 lb (1 kg) 2 lb (1 kg) 2 lb (1 kg) 8 oz (250 g)	Julienne of celery Julienne of leeks Finely sliced onions Butter	5. Sweat the leeks, celery and onion off gently in a little butter in a covered pan. This can be done while the fish is poaching.
		6. Drain the stock from the fish through a fine strainer into a suitably sized saucepan, add the liquor from the vegetables. Bring to the boil.
1 lb 4 oz (550 g) 12 oz (350 g)	Freeze 'n' flow starch Plain flour Water to mix	7. Mix the flour and starch to a smooth thin paste with cold water and stir into the fish stock. Bring back to the boil, stirring all the time, reduce the heat and simmer for 5 minutes to cook the starch.
4 oz (100 g)	Butter	8. Test the sauce for consistency and flavour, stir in the butter and correct if necessary.
		9. Arrange 8 portions of fish into each 9½ in sq deep container.
		10. Cover with the celery, leeks, onion and mushrooms.
		11. Add 2 × 15 fl. oz (1 litre) ladles of sauce.
		12. Lid and freeze for 90 minutes.
		13. Put into cold store until required.

Lid Notes
Description: Fillet of Fish Bretonne
No. of portions: 8
Oven setting: 6
Oven time: 45 minutes
Lid on
Pack weight: 2 kg

FILLET OF FISH DUGLÈRE (90–96 PORTIONS)

4 oz (100 g)	Margarine
1	Clove garlic
2 lb (1 kg)	Onions
2 stone (12·5 kg)	Fish (cod, haddock, or mock halibut)
6 tins A3	Tomatoes
4 oz (100 g)	Chopped parsley
	Seasoning
1 gal. (4·5 litres)	Fish stock or water
1½ gals. (7·5 litres)	Milk
12 oz (350 g)	Flour
1 lb 4 oz (550 g)	Waxy maize

1. Peel the garlic and rub it around the inside of the baking tins, then grease with melted margarine.

2. Finely chop the onion and sprinkle half of it into the baking tins.

3. Cut the fish into 4 oz (125 g) portions; place in the baking tins. Sprinkle with the rest of the chopped onion.

4. Drain and concasse the tomatoes, sprinkle over the fish, then sprinkle on the chopped parsley.

5. Season the fish.

6. Combine the water and milk and pour over the fish. Put into convector oven (regulo 5 for 10–15 minutes). *Do not overcook the fish.*

7. Drain the cooking liquor from the fish into a saucepan and make up the quantity to 3 gals. (12 litres). Bring to the boil.

8. Mix the flour and waxy maize to a smooth paste with cold water. Stir into the fish liquor. Bring back to the boil, reduce the heat and cook the flour, stirring all the time.

9. Check the sauce for seasoning and consistency.

10. Put 8 portions of fish into each 9½ in sq shallow container. Make sure each portion has its fair share of onion and tomato.

11. Cover with 2 × 15 fl. oz (1 litre) ladles of sauce.

12. Lid and seal.

13. Put into freezer for 90 minutes.

Lid Notes
Description: Fillet of Fish Duglère
No. of portions: 8
Oven setting: 6
Oven time: 40 minutes
Lid on
Pack weight: 2 kg

FILLET OF FISH PORTUGAISE (90–96 PORTIONS)

2 stone (12·5 kg)	Cod, haddock or mock halibut	
2 oz (50 g)	Margarine Seasoning	
8 lb (3·5 kg)	Onions	
7 cloves	Garlic	
1 pint (0·5 litre)	Cooking oil	
2 litres	White wine Seasoning	
14 tins	Tomatoes	
4 oz (100 g)	Chopped parsley	

1. Cut the fish into 4 oz (100 g) portions. Lay into buttered baking dishes and season.

2. Chop the onions and crush the garlic. Toss in the oil.

3. Add the wine and season lightly. Cook gently until the onions are softened.

4. Drain and roughly chop the tomatoes. Add to the onions and mix together until everything is evenly blended.

5. Apportion the onions and tomato over the fish.

6. Put into convector oven on regulo 5 for about 8 minutes. *Do not overcook*.

7. Put 8 portions of fish into each 9½ in sq shallow foil pack.

8. Sprinkle liberally with chopped parsley.

9. Lid and seal.

10. Put into freezer for 90 minutes.

Lid Notes
Description: Fillet of Fish Portugaise
No. of portions: 8
Oven setting: 6
Oven time: 35 minutes approx.
Lid off
Pack weight: 1·5 kg

FILLET OF FISH IN SHRIMP SAUCE
(90–96 PORTIONS)

2 stone (12·5 kg)	Fish (cod, haddock or mock halibut)	1. Cut the fish into 4 oz (125 g) portions and arrange in buttered trays.
2 oz (50 g)	Margarine	
2 lb (1 kg)	Onions	2. Finely chop the onion and sprinkle over the fish.
1½ gals. (7·5 litres)	Milk	3. Mix the milk and the fish stock together and pour over the fish.
1 gal. (4·5 litres)	Fish stock	
	Seasoning	4. Season the fish and put a bayleaf into each tray.
2	Bayleaves	
		5. Put into a convector oven for about 10 minutes on regulo 5 and cook the fish. *Do not overcook.*
		6. Drain the liquor from the fish into a saucepan and bring to the boil. Adjust to 3 gals. (12 litres).
1¼ lb (550 g)	Waxy maize	7. Mix the waxy maize and flour to a smooth paste with cold water. Stir into the boiling liquor. Bring back to the boil and reduce the heat. Allow the flour to cook gently, stirring all the time.
12 oz (350 g)	Flour	
2 oz (50 g)	Fresh parsley	8. Chop the parsley, wash it and add to the sauce.
		9. Check the sauce for seasoning and consistency.
12 tins	Shrimps	10. Drain the shrimps and apportion them evenly over the fish.
		11. Put 8 portions of fish into each 9½ in sq shallow container.
		12. Cover with 1 pint (1 litre) of sauce.
		13. Lid and seal.
		14. Put into freezer for 90 minutes.

Lid Notes
Description: Fillet of Fish in Shrimp Sauce
No. of portions: 8
Oven setting: 6
Oven time: 40 minutes
Lid on
Pack weight: 2 kg

FISH FLORENTINE (90–96 PORTIONS)

2 stone (12·5 kg)	Fish (cod, haddock or mock halibut)	1. Cut the fish into 4 oz (125 g) portions and arrange in buttered baking trays.
2 lb (1 kg)	Onions Seasoning	2. Chop the onion very finely and sprinkle over the fish. Season and cover with fish
1 gal. (4·5 litres)	Fish stock or water	stock and milk. Cook in a moderate oven
1½ gal. (7·5 litres)	Milk	(approx. 20 minutes at regulo 5). *Do not overcook.*

3. Drain the liquor from the fish into a saucepan. Adjust the quantity to 3 gals. (12 litres). Bring to the boil.

12 oz (350 g)	Flour	4. Mix the flour and freeze 'n' flow to a
1 lb 4 oz (550 g)	Freeze 'n' flow starch Water to mix	smooth paste with cold water. Stir into the boiling liquid, bring back to the boil, reduce the heat and cook the flour gently.

6 lb (2·5 kg)	Grated cheese	5. Add the cheese, mustard and garlic. Stir
2 oz (50 g)	Made mustard	until all the cheese is dissolved.
2	Crushed cloves of garlic	

6. Check the sauce for flavour and consistency.

7. Put 8 portions of fish into each foil pack.

14 lb (5 kg)	Cooked spinach	8. Put 1½ oz (50 g) of spinach on to each portion of fish.

9. Cover with 2 × 15 fl. oz (1 litre) ladles of sauce.

½ lb (250 g)	Parmesan cheese	10. Sprinkle with parmesan cheese.

11. Lid and seal.

12. Put into freezer.

Lid Notes
Description: Fish Florentine
No. of portions: 8
Oven setting: 6
Oven time: 40 minutes approx.
Lid off
Pack weight: 2 kg

FISH IN PARSLEY SAUCE (96 PORTIONS)

2 stone (12·5 kg)	Cod or haddock (fillets)
2 lb (1 kg)	Finely chopped onion Seasoning Bayleaves
2½ gals. (11·5 litres) 2 × 1 gal. (900 g)	Water Milk powder
8 oz (250 g)	Butter
1 lb 4 oz (550 g) 12 oz (350 g)	Freeze 'n' flow starch Plain flour Water to mix
4 oz (100 g) 4 tablespoons	Fresh chopped parsley Lemon juice

1. Cut the fish into 4 oz (100 g) portions and arrange in buttered baking trays.

2. Sprinkle the fish sparingly with finely chopped onion. Season and put one bayleaf into each tray.

3. Mix the milk powder and water and pour over the fish. Cover with foil and poach gently in a moderate oven.

4. Strain the liquor from the fish into a suitably sized saucepan. Add the butter and bring back to the boil.

5. Mix the flour and starch to a smooth thin paste with cold water. Stir into the fish stock and bring back to the boil stirring continuously. Reduce the heat and simmer for 5 minutes to cook the starch.

6. Stir in the parsley and lemon juice. Check for colour, flavour and consistency. Correct if necessary.

7. Place 8 portions of fish into each 9½ in sq deep foil container.

8. Cover with 2 × 15 fl. oz (1 litre) ladles of sauce.

9. Lid and seal.

10. Put into freezer for 90 minutes.

11. Store in cold room until required.

Lid Notes
Description: Fish in Parsley Sauce
No. of portions: 8
Oven setting: 6
Oven time: 45 minutes
Lid on
Pack weight: 2 kg

FISH MORNAY (90–96 PORTIONS)

2 stone (12·5 kg)	Cod, haddock or mock halibut	1. Cut the fish into 4 oz (100 g) portions and arrange in buttered baking trays.

2 lb (1 kg)   Onions
   Seasoning
1 gal. (4·5 litres)   Fish stock or water
1½ gals. (7·5 litres)   Milk

2. Chop the onion very finely and sprinkle over the fish. Season, cover with fish stock and milk. Cook in a moderate oven (approx. 20 minutes at regulo 5). *Do not overcook.*

3. Drain the liquor from the fish into a saucepan. Adjust the quantity to 3 gals. (12 litres). Bring to the boil.

12 oz (350 g)   Flour
1 lb 4 oz (550 g)   Freeze 'n' flow starch
   Water to mix

4. Mix the flour and the freeze 'n' flow to a smooth paste with cold water. Stir into the boiling liquor. Bring back to the boil, reduce the heat and cook out the flour gently.

6 lb (2·5 kg)   Grated cheese
2 oz (50 g)   Made mustard
2 cloves   Crushed garlic

5. Add the grated cheese, mustard and garlic. Stir until all the cheese is dissolved.

6. Check the sauce for flavour and consistency.

7. Place 8 portions of fish into each 9½ in sq shallow foil pack.

8. Cover with 2 × 15 fl. oz (1 litre) ladles of sauce.

½ lb (250 g)   Grated parmesan cheese

9. Sprinkle with parmesan cheese.

10. Lid and seal.

11. Put into freezer.

Lid Notes
Description: Fish Mornay
No. of portions: 8
Oven setting: 6
Oven time: 40 minutes approx.
Lid off
Pack weight: 2 kg

FISH VERONIQUE (90–96 PORTIONS)

2½ lb (1·5 kg) White grapes
1 litre White wine

2 stone (12·5 kg) Fish (cod, haddock sole* or mock halibut)

2 lb (1 kg) Onions
Seasoning
1 gal. (4·5 litres) Fish stock or water
1½ gals. (7·5 litres) Milk
Bayleaf

12 oz (350 g) Flour
1 lb 4 oz (550 g) Waxy maize
Water to mix

½ pint (0·25 litre) Cream
2 oz (50 g) Butter

1. Cut the grapes in half, remove the pips and marinade in white wine, preferably overnight.

2. Cut the fish into 4 oz (125 g) portions and arrange in buttered baking trays.

3. Very finely chop the onion and sprinkle over the fish. Season. Cover with fish stock and milk. Add a bayleaf. Cook in a moderate oven (regulo 5 for approx. 20 minutes). *Do not overcook*.

4. Drain the cooking liquor from the fish into a saucepan, bring to the boil. Adjust to 3 gals. (12 litres).

5. Meanwhile drain the wine from the grapes into another saucepan and reduce to one-third its original volume.

6. Add the wine reduction to the fish liquor.

7. Mix the flour and waxy maize to a smooth paste with cold water. Stir into the boiling fish liquor. Bring back to the boil, reduce the heat and cook out the flour, stirring all the time.

8. Stir in the cream and butter.

9. Check for flavour and consistency.

10. Put 8 portions of fish into each 9½ in sq shallow foil pack.

11. Put 6 halves of grapes onto each portion.

12. Cover with 2 × 15 fl. oz (1 litre) ladles of sauce.

13. Lid and seal.

14. Put into freezer.

* *Sole:* fillet and skin the sole and cut into 3 oz portions. Use the bones and skin to make a fish stock. Gives 140–150 portions.

Lid Notes
Description: Fish Veronique
No. of portions: 8
Oven setting: 6
Oven time: 40 minutes approx.
Lid on
Pack weight: 2 kg

FISH WALESKA STYLE (90–96 PORTIONS)

2 stone (12·5 kg)	Fish (cod, haddock, mock halibut or sole*)	1. Cut the fish into 4 oz (125 g) portions and arrange in buttered baking trays.
2 lb (1 kg)	Onions Seasoning	2. Very finely chop the onion and sprinkle over the fish. Season, cover with the milk and fish stock. Add a bayleaf and cook in a moderate oven for 20 minutes at regulo 5 in convector oven. *Do not overcook*.
1 gal. (4·5 litres) 1½ gals. (7·5 litres)	Fish stock or water Milk Bayleaf	
3 lb (1·5 kg)	Mushrooms	3. Wash and drain the mushrooms. Slice thinly.
½ lb (250 g) ½ pint (250 cc)	Margarine Water Seasoning	4. Place the mushrooms in a pan with the margarine and water. Season, cover with a lid and gently cook the mushrooms.
		5. Drain the cooking liquor from the fish into a saucepan. Add the liquor from the mushrooms. Adjust the quantity to 14 pints (8 litres) and bring to the boil.
12 oz (350 g) 1 lb (550 g)	Flour Waxy maize Water to mix	6. Mix the flour and waxy maize to a smooth paste with cold water. Stir into boiling liquor. Bring back to the boil and cook the flour gently, stirring all the time.
6 lb (2·5 kg) 3 1½ oz (50 g) 1 pint (0·5 litre) 2 oz (50 g)	Grated cheese Crushed cloves garlic Mustard Cream Butter	7. Add the grated cheese, garlic and mustard. Stir until the cheese is dissolved. Add cream and butter.
		8. Check the sauce for flavour and consistency.
		9. Put 8 portions of fish into each 9½ in sq shallow foil pack.
		10. Put 1 dessertspoonful of mushrooms onto each portion.
2 lb (1 kg)	Freeflow prawns	11. Put 1 dessertspoonful of prawns onto each portion.
		12. Cover with one 20 fl. oz (0·5 litre) ladle of sauce.
½ lb (250 g)	Grated parmesan cheese	13. Sprinkle with parmesan cheese.
		14. Lid and seal.
		15. Place into freezer.

* Sole should be filleted and skinned. Use the bones and skin to make a fish stock. Cut into 3 oz portions giving 140–150 portions.

Lid Notes
Description: Fish Waleska Style
No. of portions: 8
Oven setting: 6
Oven time: 40 minutes approx.
Lid off, if served from pack
Lid on, if served from flat
Pack weight: 2 kg

FRIED FISH IN BATTER (90–96 PORTIONS)

2 stones (12·5 kg)	Fish (cod, haddock or mock halibut)	1. Cut the fish into 4 oz (110 g) portions.
1½ lb (700 g)	Seasoned flour	2. Pass the fish through seasoned flour, shaking off any excess flour.
4 pints (2·25 litres)	Batter*	

3. Pass the floured fish through the batter between thumb and forefinger.

4. Drop the battered fish immediately into hot clean fat (340°F) and set the batter. Do not allow the fish to colour.

5. Drain on wire racks.

6. Place on to foil covered shelves of the freezing trolleys.

7. Cover the fish with foil, place on to the trolleys and put into the freezer.

8. Freeze for 90 minutes. Remove the fish from the tunnel and pack in polythene bags, 25 portions to the bag.

* *Batter*
  3 lb Batter mix
  3¾ pints Water

Label on Bags
Description: Fried Fish in Batter
Deep fry at 350°F for 5–7 minutes

FRIED FISH IN BREADCRUMBS (90–96 PORTIONS)

2 stone (12·5 kg)	Fish (cod, haddock or mock halibut)	1. Cut the fish into 4 oz (110 g) portions.
1½ lb (700 g)	Seasoned flour	2. Pass the fish through seasoned flour, shake off any excess.
3 pints (1·75 litres)	Egg wash*	3. Pass the floured fish through the egg wash, removing any excess between thumb and forefinger.
8 lb (3·5 kg)	Breadcrumbs	4. Pass the fish through the breadcrumbs, making sure the fish is well covered. Shake off excess crumbs.
		5. Place the breadcrumbed fish on to foil covered trolley shelves. Do not overlap.
		6. Cover the fish with foil and put on to the trolley.
		7. Freeze for 90 minutes. Remove from the tunnel and pack the fish in polythene bags, 25 portions to the bag.

* *Egg Wash*
9 Eggs
1·75 litres Milk

Label on Bags
Description: Fried Fish in Breadcrumbs
Deep fry at 350°F for 5–7 minutes

GOUJONS OF SOLE (96 PORTIONS)

2 stone (12·5 kg)	Lemon sole
1 lb (500 g)	Seasoned flour
12 2 pints (1 litre)	Eggs⎱ Milk⎰ Egg wash
3 lb (1·5 kg)	Breadcrumbs

1. Skin and fillet the sole. Cut into strips the size of gudgeon, about $2\frac{1}{2}$ in × $\frac{1}{2}$ in.
2. Toss in seasoned flour, shake off excess flour.
3. Pass through the egg wash. Allow to drain a little.
4. Pass through the breadcrumbs, making sure that all the fish is coated. Gently shake off any loose breadcrumbs.
5. Weigh $1\frac{1}{2}$ lb (1 kg) of goujons into each $9\frac{1}{2}$ in sq shallow container.
6. Lid and seal.
7. Put into freezer for 90 minutes.
8. Store in holding fridge until required.

Lid Notes
Description: Goujons of Sole
No. of portions: 8
Deep fry at 350°F for 4–5 minutes

SALMON MOUSSE (110 PORTIONS)

1 litre	Red wine	1. Make up a bouillon with the wine, vinegar, peppercorns, bayleaves, lemons and salt and water in a suitable pan.
½ pint (250 cc)	Vinegar	
2	Bayleaves	
4	Lemons (sliced)	
Few	Peppercorns	
	Salt	
	Water	
12 lb (5·5 kg)	Salmon	2. Defrost the salmon and poach it gently in the bouillon.
		3. When the salmon is cooked, remove it carefully from the bouillon. Allow it to cool a little then remove the skin. Carefully take out all the bones and put the flesh into a mixing bowl and beat it until it is smooth.
8 oz (250 g)	Butter	4. Make a roux with the butter and flour and cook it gently without colouring.
8 oz (250 g)	Flour	
4 pints (2·25 litres)	Milk	5. Gradually beat in the milk and add the onion clouté. Simmer the sauce until the flour is cooked. Allow it to cool.
1	Onion studded with cloves	
4 pints (2·25 litres)	Fish stock	6. Measure 4 pints (2·25 litres) from the fish stock and dissolve the gelatine in it. Pass through a muslin into a clean bowl and allow it to cool until it is on the point of setting.
2 oz (50 g)	Gelatine	
3 pints (1·75 litre)	Cream	7. Whip the cream until it just drops off a spoon.
12	Egg whites	8. Stiffly beat egg whites.
	Colouring	9. Mix the white sauce with the fish, add the jelly and fold in the cream. Taste for seasoning and correct the colour if necessary.
	Seasoning	
		10. Fold in the egg whites.
		11. Measure 4 oz (125 g) of the mixture into each individual smooth sided plastic container.
		12. Seal with suitably labelled cling film.
		13. Put into freezer for 90 minutes.
		14. Store under refrigeration until required.

Lid Notes
Description: Salmon Mousse
Defrost and serve

SCAMPI AMERICAN STYLE (80 PORTIONS)

20 lb (9 kg)	Jumbo scampi	1. Melt the butter in a saucepan and sweat off the finely chopped onions and crushed garlic under a lid and without colouring.
1 lb (500 g)	Butter	
1 clove	Crushed garlic	
2½ lb (1·35 kg)	Chopped onions	
		2. Add the scampi and allow to cook gently for about 10 minutes.
3 gills (425 cc)	Brandy	3. Add the brandy and reduce it.
1 lb (500 g)	Tomato puree	4. Add the tomato puree and concassé tomatoes. Simmer until the scampi is cooked.
9 lb (4 kg)	Concassé tomatoes	
6 pints (3·5 litres)	Dry white wine	5. Reduce the wine and add it to the sauce.
		6. Weigh 2 lb (900 g) scampi into each 9½ in sq shallow container.
		7. Mask with one 15 oz (425 cc) ladle of sauce.
2 oz (50 g)	Finely chopped parsley	8. Sprinkle with freshly chopped parsley.
		9. Lid and seal.
		10. Put into freezer for 90 minutes.
		11. Store in holding fridge until required.

Lid Notes
Description: Scampi American Style
No. of portions: 8
Oven setting: 5
Oven time: 40 minutes
Lid on
Pack weight: 1·325 kg

SOLE DIEPPOISE (96 PORTIONS)

2 stone (12·5 kg) (3–4 oz) (125 g)	Lemon sole (fillets)

1. Skin the dark fillets and trim the white fillets. Fold and arrange in buttered baking trays.

2. Put the skins and trimmings into a saucepan and cover with water. Bring to the boil, reduce the heat and simmer to make a clear fish stock.

2 lb (1 kg)	Finely chopped onion Pepper and salt
1 gal. (4·5 litres)	Fish stock
1½ gals. (7·5 litres)	Milk
	Bayleaf

3. Sprinkle the fish with finely chopped onion, season, cover with fish stock and milk. Add a bayleaf to each tray. Cook in a moderate oven, approx. 20 minutes on regulo 5. *Do not overcook.*

2 lb (1 kg)	Mushrooms (button)

4. Wash and drain the mushrooms.

½ lb (50 g)	Margarine
½ pint (25 cc)	Water
	Seasoning

5. Slice the mushrooms thinly and place into a saucepan with the margarine and water. Season, cover with a lid and cook gently.

6. Drain the cooking liquor from the fish into a saucepan. Adjust the quantity to 14 pints (8 litres). Bring to the boil, reduce the heat and simmer until all the proteins have coagulated.

7. Strain into a clean saucepan.

12 oz (350 g)	Flour
1 lb 4 oz (550 g)	Freeze 'n' flow starch

8. Mix the flour and freeze 'n' flow to a smooth paste with cold water. Stir into the boiling liquor. reduce the heat and gently cook the flour, stirring all the time.

½ litre	Dry white wine

9. Reduce the wine over heat to half its volume, add to the sauce.

1 pint (0·5 litre)	Cream
2 oz (50 g)	Butter

10. Stir in the cream and the butter. Check the sauce for consistency and flavour.

2 lb (1 kg)	Free-flow prawns

11. Defrost the prawns.

1 gal. (4·5 litres)	Mussels (in brine)

12. Wash and drain the mussels.

13. Arrange 8 portions of sole into each 9¼ in sq shallow foil container.

14. Put one dessertspoon of mushrooms, 1 dessertspoon of prawns and one dessertspoon of mussels on to each portion.

15. Cover with 1½ pints (1 litre) of sauce.

16. Lid and seal.

17. Put into blast-freeze tunnel for 90 minutes.

Lid Notes
Description: Sole Dieppoise
No. of portions: 8
Oven setting: 6
Oven time: 40 minutes
Lid on
Pack weight: 2 kg

Farinaceous and Vegetarian

CHEESE AND NUT PANCAKES (24 PORTIONS)

1¾ lb (800 g)	Yorkshire pudding mix
2½ pints (1·45 litres)	Water
2 lb (1 kg)	Grated cheese
2 lb (1 kg)	Mixed nuts
½ pint (250 cc)	Nut or corn oil

1. Mix the water and Yorkshire pudding mix together following manufacturers instructions. Let it stand for 10 minutes.

2. Chop the mixed nuts and mix them with the grated cheese.

3. Cover the bottom of an iron frying pan (9 in diameter) with nut oil and heat until a blue haze comes from it. Pour off any excess fat.

4. Ladle sufficient batter mix into the pan to just cover the bottom, this is best achieved by holding the pan at a 45° angle and pouring the batter down from the top; twist the pan until the surface is covered and pour back any batter that does not adhere to the pan.

5. Brown the pancake on one side then turn it over or toss it and brown the other side.

6. Stack the pancakes one on top of the other.

7. Put 1⅓ oz (35 g) of the cheese and nut mix into the centre of each pancake. Roll it up and pack 6 into a 9½ in sq shallow container.

8. Lid and seal.

9. Put into freezing tunnel for 90 minutes.

10. Store in holding fridge until required.

Lid Notes
Description: Cheese and Nut Pancakes
No. of portions: 3
Oven setting: 6
Oven time: 35 minutes
Lid on
Pack weight: 750 g

CHEESE AND ONION PASTIES (69 PORTIONS)

2½ lb (1 kg)	Finely chopped onions	1. Just cover the onions with water and cook with a lid on the stove.
5 lb (2·25 kg)	Cheese	2. Cut the cheese into ¼ in cubes.
5 lb (2·25 kg)	Cooked potatoes	3. Cut the cooked potatoes into ¼ in cubes.
	Seasoning	4. Mix the cheese, potatoes and onions, together with the water that the onions were cooked in, in a suitably sized saucepan. Season to taste.
10 lb (4·5 kg)	Shortcrust pastry	5. Roll the pastry and cut into 6½ in diameter circles. These should weigh about 4 oz (110 g) each.
½ pint (250 cc)	Egg wash	6. Brush the outsides of the pastry with egg wash.
		7. Weigh 3 oz (85 g) of the cheese filling onto each pastry circle.
		8. Fold the pastry over to form a pasty. Seal the edge and crinkle.
		9. Egg wash.
		10. Completely cover each pasty with foil or cling film.
		11. Put into the tunnel and freeze.
		12. Pack into boxes and label.

Label
Oven setting: 5
Oven time: 35 minutes
Pack weight: 195 g

CHEESE, POTATO AND ONION PIE
(240 PORTIONS)

15 lb (7 kg)	Onions	1. Peel and chop the onions. Sweat them in just sufficient water to moisten them.
30 lb (14 kg)	Potatoes	2. Par-boil the potatoes in salted water and cut into $\frac{3}{8}$ in cubes.
20 lb (10 kg)	Cheese	3. Cut the cheese into $\frac{3}{8}$ in cubes.
	Seasoning	4. Combine the onions, potatoes and cheese in a large bowl and season.
20 lb (10 kg)	Shortcrust pastry	5. Line the 9 in diameter pie foils with 8 oz (250 g) pastry.
		6. Add $1\frac{1}{2}$ lb (625 g) cheese and potato filling.
		7. Top with 7 oz (200 g) pastry.
$\frac{1}{2}$ pint (250 cc)	Egg wash	8. Brush with egg wash.
		9. Cover with suitably labelled foil.
		10. Put into freezer for 90 minutes.
		11. Store in holding fridge until required.

Lid Notes
Description: Cheese, Potato and Onion Pie
No. of portions: 6
Oven setting: 5
Oven time: 35 minutes
Lid off
Pack weight: 1·135 kg

EGG CROQUETTES (72 PORTIONS)

52	Eggs	1. Hard boil the eggs, plunge into cold water and shell them.
1 gal. (4·5 litres)	Milk	2. Infuse the onion, peppercorns and bayleaf in the milk and bring slowly to the boil.
1 lb (500 g)	Onions	
10	Peppercorns	
1	Bayleaf	
1½ lb (750 g)	Margarine	3. Melt the margarine in a saucepan, stir in the flour and cook gently without colouring.
1½ lb (750 g)	Flour	
	Seasoning	
		4. Gradually stir in the milk and bring back slowly to the boil. Cook the panade.
		5. Season and allow the panade to cool.
12	Eggs	6. Mince the hard boiled eggs, stir into the panade. Beat in raw eggs to bind. Check the seasoning.
		7. With a size 26 ice cream scoop portion the egg mixture into 2 oz (50 g) portions.
1 lb (500 g)	Seasoned flour	8. Pass through seasoned flour.
3 pints (1·5 litres)	Egg wash	9. Pass through egg wash.
1¾ lb (1 kg)	Breadcrumbs	10. Pass through breadcrumbs.
		11. Reshape by agitating in the ice cream scoop.
		12. Pack 16 croquettes in each 9½ in sq shallow container.
		13. Lid and seal.
		14. Put into freezer for 90 minutes.
		15. Store in holding fridge until required.

Lid Notes
Description: Egg Croquettes
No. of portions: 8
Deep fry
Pack weight: 1 kg

EGG FLORENTINE (90 PORTIONS)

24 lb (11 kg)	Leaf spinach (frozen)	1. Cook the spinach in plenty of salted water. Drain well.
1 lb (500 g) 1 lb (500 g)	Margarine Flour	2. Make a roux with the margarine and flour.
2½ gals. (12 litres) 1	Milk Onion studded with cloves	3. Infuse the onion clouté in the milk and bring to the boil.
	Seasoning	4. Combine the milk and the roux to make a sauce. Season and simmer for 5 minutes. This sauce will be thin.
2½ lb (1·5 kg) 1 tablespoon 1 clove	Grated cheese Made mustard Crushed garlic	5. Stir the grated cheese into the sauce. Add the mustard and the crushed clove of garlic.
1 lb 4 oz (650 g)	Waxy maize	6. Mix the waxy maize to a smooth creamy paste with cold water. Stir into the sauce to correct the consistency.
		7. Check the sauce for seasoning.
		8. Weigh 1 lb (500 g) of cooked spinach into each 9½ in sq shallow pack. Spread it evenly across the bottom of the pack and make 6 indentations in it with a ladle (7 oz) or rolling pin.
90	Eggs	9. Crack an egg into each indentation.
		10. Cover the whole with 2 × 15 oz (850 cc) ladles of cheese sauce.
8 oz (250 g)	Parmesan cheese	11. Sprinkle with parmesan cheese.
		12. Lid and seal.
		13. Place into the freezer tunnel for 90 minutes.

Lid Notes
Description: Egg Florentine
No. of portions: 6
Oven setting: 6
Oven time: 40 minutes
Lid off
Pack weight: 1·5 kg

GNOCCHI GRATINÉE (96 PORTIONS)

3 lb (1·5 kg)	Margarine	1. Put the margarine, milk, salt and nutmeg into a pan and bring to the boil.
12 pints (7 litres)	Milk	
1 tablespoon	Salt	
2 teaspoons	Grated nutmeg	

6 lb (3 kg)   Flour

2. Stir in the flour and beat until the mixture is stiff and forms a ball. Remove from the heat and allow to cool a little.

6 dozen (approx.)   Eggs

3. Put the mixture into a mixing bowl and beat in the eggs a few at a time until the mixture is smooth and just holds its shape when lifted with a spoon.

4. Half fill a bratt pan with salted water and bring to the boil. Reduce heat to a fast simmer.

5. Stretch a cheese wire across the bratt pan.

6. Put the mixture into a forcing bag containing a $\frac{1}{2}$ in plain nozzle.

7. Pipe the mixture across the wire so that it falls into the bratt pan in one-inch lengths.

8. Cook the gnocchi until it floats to the surface. Remove from the bratt pan with a perforated scoop and drain thoroughly.

3 lb (1·5 kg)   Margarine

9. Melt the margarine in a bratt pan and toss the gnocchi into it. Take care not to break the gnocchi and make sure that each piece is coated with margarine.

10. Weigh 2 lb (1 kg) of gnocchi into each $9\frac{1}{2}$ in sq shallow container.

6 lb (3 kg)   Grated cheese

11. Sprinkle 6 oz (175 g) grated cheese on to each container.

12. Lid and seal.

13. Put into freezer for 90 minutes.

14. Store in holding fridge until required.

Lid Notes
Description: Gnocchi Gratinée
No. of portions: 6
Oven setting: 5
Oven time: 40 minutes
Lid off
Pack weight: 1·175 kg

GNOCCHI IN TOMATO SAUCE (96 PORTIONS)

6 lb (3 kg)	Flour	
3 lb (1·5 kg)	Margarine	
12 pints (7 litres)	Milk	
1 tablespoon	Salt	
6 dozen (approx.)	Eggs	

1. Make up the paste as in Gnocchi Gratinee. Cook and dry the paste but do not toss it in melted margarine.

1 lb (500 g)	Bacon dripping
2 lb (1 kg)	Bacon scraps
or	
2 lb	Margarine

2. Melt the dripping in a saucepan and fry the bacon scraps.
   Note: If the Gnocchi in Tomato Sauce is to be featured in vegetarian menus omit the dripping and bacon scraps and substitute margarine.

1½ lb (750 g)	Chopped onions
1½ lb (750 g)	Chopped carrots
½ lb (250 g)	Chopped celery
2 cloves	Crushed garlic

3. Add the onions, carrots, celery and garlic. Sweat under a lid without colouring.

1 lb (500 g)	Flour
1 teaspoon	Thyme
2 lb (1 kg)	Tomato puree

4. Stir in the flour and cook it. Add the thyme and tomato puree.

2 gals. (9 litres)	Water or stock
4 oz (100 g)	Sugar
	Seasoning

5. Stir in the water or stock add the sugar and season to taste. Bring to the boil, stirring to avoid lumps. Reduce the heat and simmer gently until the sauce is cooked.

6. Strain the sauce into a clean saucepan and remove excess fat.

1 lb (500 g)	Freeze 'n' flow starch

7. Reconstitute the freeze 'n' flow and stir into the sauce. Reduce the heat and simmer until the sauce is cooked.

8. Check the sauce for seasoning, colour and consistency, adjust if necessary.

9. Weigh 2 lb (1 kg) of gnocchi into each 9½ in sq shallow container.

10. Add 1 pint (0·5 litre) of tomato sauce (1 lb 4 oz).

11. Lid and seal.

12. Put into freezer for 90 minutes.

13. Store in holding fridge until required.

Lid Notes
Description: Gnocchi in Tomato Sauce
No. of portions: 6
Oven setting: 5
Oven time: 45 minutes
Lid on
Pack weight: 1·5 kg

MACARONI AU GRATIN (160 PORTIONS)

6 lb (3 kg)	Macaroni Boiling water Salt	1. Cook the macaroni in plenty of salted water. Drain and refresh. This can be done the day before it is required if necessary.
3 gals. (13·5 litres) ½ lb (250 g)	Milk Margarine	2. Bring the milk and margarine to the boil in a saucepan.
1 lb (500 g) 1 lb 4 oz (650 g)	Flour Freeze 'n' flow starch	3. Mix the flour and freeze 'n' flow to a smooth paste with cold water. Stir it into the boiling milk. Bring back to the boil and simmer until the flour is cooked. Stir occasionally to avoid burning.
3 oz (75 g) 3 cloves	Made mustard Crushed garlic Pepper and salt	4. Season the sauce and add the mustard and garlic.
7½ lb (3·5 kg)	Grated cheese	5. Add the grated cheese and stir until the sauce is smooth. Check for colour, seasoning and consistency.
		6. Add the macaroni to the sauce and stir well.
		7. Weigh 3 lb 2 oz (1·5 kg) macaroni into each 9½ in sq shallow container.
8 oz (250 g)	Parmesan cheese	8. Sprinkle with grated parmesan cheese.
		9. Lid and seal.
		10. Put into freezer for 90 minutes.
		11. Store in holding fridge until required.

Lid Notes
Description: Macaroni au Gratin
No. of portions: 8
Oven setting: 5
Oven time: 45 minutes
Lid off
Pack weight: 1·5 kg

PANHAGGERTY (180 PORTIONS)

15 lb (7 kg)	Onions	1. Peel and chop the onions. Put them into a saucepan with a little salt and water. Cook them with the lid on.
30 lb (14 kg)	Peeled potatoes	2. Par-boil the potatoes in salted water. Drain, cool and cut into $\frac{1}{2}$ in cubes.
20 lb (9·5 kg)	Cheese	3. Cut the cheese into $\frac{1}{2}$ in cubes.
	Seasoning	4. Combine all the ingredients and season with pepper and salt.
		5. Weigh 3 lb (1·5 kg) of the mixture into each $9\frac{1}{2}$ in sq shallow container.
4 pints (2·25 litres)	Milk	6. Add 4 oz (75 cc) milk to each pack.
		7. Lid and seal.
		8. Put into freezer for 90 minutes.
		9. Store under refrigeration until required.

Lid Notes
Description: Panhaggerty
No. of portions: 6
Oven setting: 6
Oven time: 40 minutes
Lid off
Pack weight: 1·5 kg

PIZZA (240 PORTIONS)

4 oz (125 g)	Dried yeast	1. Warm the water to blood heat; add the yeast and sugar and dissolve.
4 oz (125 g)	Sugar	
2 pints (1 litre)	Water	
12 lb (5·5 kg)	Plain flour	2. Sieve the flour into a mixing bowl and add the salt and eggs.
1 tablespoon	Salt	
½ doz.	Eggs	
8–9 pints (4·5–5 litres)	Water	3. Add the yeast mixture and the water to form a very soft dough. Beat with the dough hook until the batter is very smooth and elastic. The dough should stick to a metal spoon then slowly fall like a thick batter.

4. Cover with a damp cloth and put into a warm place to prove until the dough has doubled in size.

8 lb (3·5 kg)	Onions	5. Meanwhile clean and slice the onions and mushrooms. Sweat them in a little water and the margarine in a tightly lidded pan. Season them to taste and allow to cool.
3 lb (1·5 kg)	Mushrooms	
	Seasoning	
8 oz (250 g)	Margarine	
3 × A10	Tomatoes	6. Strain the juice from the tomatoes into a saucepan. Mince the tomatoes and add to the juice. Add the herbs and pepper and salt and bring to the boil.
1 teaspoon	Mixed herbs	
	Seasoning	
6 oz (200 g)	Freeze 'n' flow starch	7. Reconstitute the freeze 'n' flow with cold water and stir into the tomatoes. Bring back to the boil, reduce the heat and simmer until the starch is cooked.

8. Weigh 6 oz (175 g) dough into each foil plate and smooth it around the inside of the foil.

9. Add 5 oz (150 g) tomato topping and 3 oz (85 g) of mushroom and onion.

15 lb (7 kg)	Grated cheese	10. Sprinkle the top with 4 oz (115 g) grated cheese.

11. Cover with suitably labelled foil.

12. Put onto the freezer trolleys in a warm place and allow them to prove for 20–25 minutes.

13. Put into freezer for 90 minutes.

14. Store under refrigeration until required.

Lid Notes
Description: Pizza
No. of portions: 4
Oven setting: 4
Oven time: 35 minutes
Lid off
Pack weight: 525 g

RATATOUILLE (180 PORTIONS)

16 lb (7 kg)	Peppers
10 lb (4·5 kg)	Onions
9 lb (4 kg)	Mushrooms
20 lb (9 kg)	Aubergines
5 lb (2 kg)	Green beans
25 lb (11 kg)	Tomatoes
4 pints (2 litres)	Olive oil
3 cloves	Crushed garlic
1 teaspoon	Mixed herbs
	Seasoning
2 oz (50 g)	Chopped parsley

1. Clean and prepare all the vegetables.

2. Heat the olive oil in a bratt pan and gently fry the onions.

3. Add the garlic, herbs, parsley and seasoning and the sliced green beans.

4. Add the mushrooms, peppers and aubergines and finally the tomatoes. Cover with a lid and lightly cook for a few minutes until the vegetables are slightly undercooked. Check for seasoning.

5. Weigh $2\frac{3}{4}$ lb (1·25 kg) of ratatouille into each $9\frac{1}{2}$ in sq shallow container.

6. Lid and seal.

7. Put into freezer for 90 minutes.

8. Store in holding fridge until required.

Lid Notes
Description: Ratatouille
No. of portions: 6
Oven setting: 6
Oven time: 35 minutes
Lid on
Pack weight: 1·25 kg

RISOTTO (90 PORTIONS)

5 lb (2 kg)	Onions
2½ lb (1 kg)	Mushrooms
3 lb (1·5 kg)	Green peppers
1 pint (500 cc)	Olive oil
10 lb (4·5 kg)	Rice
1 gal. (4·5 litres)	Water
	Salt to taste
	Cayenne pepper
2 × A2	Red pimentoes

1. Clean and slice the onions, mushrooms and peppers.

2. Sweat the vegetables without colouring in a 7½ gal. saucepan.

3. Add the rice and stir well so that each grain of rice is coated in oil.

4. Stir in the water and season to taste. Bring slowly up to the boil, stirring frequently to stop the rice from congealing.

5. Place a lid on the saucepan and put it into a pre-heated oven on regulo 5.

6. The rice will absorb all the water. Remove from the oven just before the rice is properly cooked out.

7. Drain and chop the pimentoes and stir into the risotto.

8. Weigh 2½ lb (1 kg) into each 9½ in sq shallow container.

9. Lid and seal.

10. Freeze for 90 minutes.

11. Store in holding fridge until required.

Lid Notes
Description: Risotto
No. of portions: 6
Oven setting: 5
Oven time: 40 minutes
Lid on
Pack weight: 1 kg

SPAGHETTI AU GRATIN (160 PORTIONS)

8 lb (3·5 kg)	Spaghetti Boiling water Salt	1. Cook the spaghetti in plenty of salted water. Drain and refresh. This can be done the day before it is required if necessary.
3 gals. (13·5 litres) ½ lb (250 g)	Milk Margarine	2. Bring the milk and margarine to the boil in a saucepan.
1 lb (500 g) 1 lb 4 oz (650 g)	Flour Freeze 'n' flow starch	3. Mix the flour and freeze 'n' flow to a smooth paste with cold water. Stir it into the boiling milk. Bring back to the boil and simmer until the flour is cooked. Stir occasionally to avoid burning.
3 oz (75 g) 3 cloves	Made mustard Crushed garlic ᵢ Pepper and salt	4. Season the sauce and add the mustard and garlic.
7½ lb (3 kg)	Grated cheese	5. Add the grated cheese and stir until the sauce is smooth. Check for colour, seasoning and consistency.
		6. Add the spaghetti to the sauce and stir well.
		7. Weigh 3 lb 2 oz (1·5 kg) spaghetti into each 9½ in sq shallow container.
8 oz (250 g)	Parmesan cheese	8. Sprinkle with grated parmesan cheese.
		9. Lid and seal.
		10. Put into freezer for 90 minutes.
		11. Store in holding fridge until required.

Lid Notes
Description: Spaghetti au Gratin
No. of portions: 8
Oven setting: 5
Oven time: 45 minutes
Lid off
Pack weight: 1·5 kg

STUFFED AUBERGINES (70 PORTIONS)

2 boxes		
30 lb (13·5 kg)	Aubergines	1. Remove stalk from and wash the aubergines.
	Boiling salted water	2. Cut the aubergines in half lengthwise and put in a large pan containing plenty of boiling salted water.
		3. Bring the pan back to the boil, simmer for 2 minutes, drain the aubergines and refresh in cold water. Drain again.
		4. Scoop out the middle of the aubergines leaving a $\frac{1}{4}$ in layer all around the skin. Keep the scooped out centres on one side for use later on.
$3\frac{1}{2}$ lb (1·5 kg)	Chopped onions	5. Sweat off the onions, mushrooms and green peppers in the vegetable oil.
1 pint (500 cc)	Vegetable oil	
$1\frac{1}{2}$ lb (1 kg)	Chopped mushrooms	
2 lb (1 kg)	Chopped green peppers	
7 lb (3 kg)	Brown rice	6. Add the brown rice and stir well making sure that the grains of rice are coated in oil.
$5\frac{1}{4}$ pints (3 litres)	Water	7. Add the water and seasoning. Stir well and bring slowly to the boil, stirring frequently to keep the grains of rice separate.
	Salt	
	Cayenne pepper	
		8. Put a lid on the pan and place it in a hot oven, regulo 5. Stir regularly.
		9. Just before it is cooked remove it from the oven. Chop the flesh from the aubergines and add it to the risotto.
		10. Put 5 halves of aubergine into each $9\frac{1}{2}$ in sq deep foil pack. Fill each half with risotto.
		11. Lid and seal.
		12. Put into freezing tunnel for 90 minutes.
		13. Store in holding fridge until required.

Lid Notes
Description: Stuffed Aubergines
No. of portions: 5
Oven setting: 5
Oven time: 40 minutes
Lid on
Pack weight: 1 kg

STUFFED MARROW (50 PORTIONS)

$2\frac{1}{4}$ lb (1 kg)   Chopped onion
1 lb (500 g)   Chopped mushrooms
1 lb (500 g)   Margarine
2 cloves   Crushed garlic

$1\frac{1}{2}$ lb (750 g)   Rice

$1\frac{1}{5}$ pint (650 cc)   Water
Pepper and salt

$1\frac{1}{2}$ lb (750 g)   Concassé tomatoes
1 lb (500 g)   Grated cheese

12 lb (5·5 kg)   Marrows

Boiling salted water

1. Sweat the onions, mushrooms and garlic in the margarine in a saucepan.

2. Stir in the rice and make sure that each grain of rice is coated in margarine.

3. Add the water, season to taste and stirring regularly bring to the boil.

4. Cover the pan with a lid and finish cooking the rice in the oven on regulo 5.

5. Just before the rice is cooked stir in tomatoes and cheese. Return the pan to the oven for about 5 minutes to cook the tomatoes. Check the rice mixture for seasoning and texture. It should now be quite dry, like a risotto.

6. While the rice is cooking peel the marrows and remove the pips.

7. Cut the marrow into 3 oz (85 g) pieces.

8. Cook the marrow in boiling salted water. Do not overcook. Drain, refresh and drain again.

9. Put 6 pieces of marrow into each $9\frac{1}{2}$ in sq shallow container.

10. Put 2 × No. 26 scoops of risotto into each portion of marrow.

11. Lid and seal.

12. Put into the freezer for 90 minutes.

13. Store in holding fridge until required.

Lid Notes
Description: Stuffed Marrow
No. of portions: 6
Oven setting: 5
Oven time: 40 minutes
Lid on
Pack weight: 1·25 kg

SWEETCORN CROQUETTES (136 PORTIONS)

4½ lb (2 kg)	Porridge oats
7 pints (4 litres)	Water
	Seasoning
3 × A10	Sweetcorn
1 lb (450 g)	Brazil nuts
1 lb (450 g)	Hazel nuts
1 lb (450 g)	Peanuts
3 packets	Savourmix
	Seasoned flour
	Egg wash
	Breadcrumbs

1. Bring the water to the boil, stir in the porridge oats and season. Cook until the mixture is very thick—like a panade.

2. Open and drain the sweetcorn.

3. Put the nuts through the mincer, using the plate with the largest holes.

4. Put the cooked porridge oats into a mixing bowl, add the sweetcorn and the nuts.

5. Add the savourmix gradually until the mixture forms a stiff dough that can be shaped into croquettes.

6. Check the mixture for seasoning.

7. Using size No. 24 ice cream scoop, portion the mixture into 2 oz (50 g) portions.

8. Pass through seasoned flour.

9. Egg wash and breadcrumb them and reshape if necessary.

10. Pack 16 croquettes (2 to a portion) into each 9½ in sq foil container.

11. Lid and seal.

12. Put into freezer for 90 minutes.

13. Store in holding fridge until required.

Lid Notes
Description: Sweetcorn Croquettes
No. of portions: 8
Deep fry
Pack weight: 1 kg

Sweets

APPLE TART (320 PORTIONS)

6 × A10	Solid pack apples	1. Slice the apples on the slicing attachment into a suitably sized bowl.
6 pints (3·5 litres) 6 lb (2·75 kg)	Water Sugar Pinch powdered cloves	2. Put the water and sugar into a saucepan and bring to the boil. Add the cloves.
9 oz (255 g)	Freeze 'n' flow starch	3. Reconstitute the freeze 'n' flow with cold water and stir into the syrup. Bring back to the boil, reduce the heat to a simmer and cook the starch.
		4. Pour the thickened syrup on to the sliced apples and mix them together well and allow to cool.
20 lb (10 kg)	Sweet shortcrust pastry	5. Line the 9 in foil plates with 8 oz (250 g) pastry.
		6. Add 1 lb 2 oz (500 g) apple pie filling.
		7. Cover with 7 oz (200 g) pastry.
½ pint (250 cc)	Egg wash	8. Brush with egg wash.
	Sugar to dredge	9. Sprinkle with sugar.
		10. Wrap with suitably labelled foil.
		11. Put into freezer for 90 minutes.
		12. Store in holding fridge until required.

Lid Notes
Description: Apple Tart
No. of portions: 8
Oven setting: 5
Oven time: 35 minutes
Lid off
Pack weight: 1 kg

APPLE AND BLACKBERRY TART (320 PORTIONS)

2 × A10	Blackberries	1. Drain the juice from the blackberries into a saucepan and make it up to 6 pints (3·5 litres) with water.
6 lb (2·75 kg)	Sugar	2. Add the sugar to the juice and bring to the boil.
6 oz (170 g)	Freeze 'n' flow starch	3. Reconstitute the freeze 'n' flow with cold water and stir into the syrup. Bring back to the boil, reduce the heat and simmer until the sauce clears.
4 × A10	Solid pack apples	4. Slice the apples on the slicing attachment into a suitably sized container. Add the blackberries and the sauce. Stir well to blend the fruit and sauce properly. Allow to cool.
20 lb (10 kg)	Sweet shortcrust pastry	5. Line the 9 in diameter pie foils with 8 oz (250 g) pastry.
		6. Add 1 lb 2 oz (550 g) pie filling.
		7. Cover with 7 oz (200 g) pastry.
½ pint (250 cc)	Egg wash	8. Brush with egg wash.
	Sugar to dredge	9. Sprinkle with sugar.
		10. Wrap with suitably labelled foil sheets.
		11. Put into freezer for 90 minutes.
		12. Store in holding fridge until required.

Lid Notes
Description: Apple and Blackberry Tart
No. of portions: 8
Oven setting: 5
Oven time: 35 minutes
Lid off
Pack weight: 1 kg

APPLE AND BLACKCURRANT PIE (320 PORTIONS)

2 × A10 6 lb (2·75 kg)	Blackcurrants Sugar Cold water	1. Drain the juice of the blackcurrants into a saucepan. Make up to 6 pints with water. Add the sugar. Bring to the boil.
9 oz (255 g)	Freeze 'n' flow starch	2. Reconstitute the freeze 'n' flow with cold water. Stir into the boiling syrup. Bring back to the boil, reduce the heat and simmer until the mixture clears, stirring all the time. Allow to cool.
6 × A10	Solid pack apples	3. Slice the apples on the slicing attachment into a suitably sized bowl.
		4. Add the blackcurrants and the thickened syrup to the apples and mix well together.
20 lb (10 kg)	Sweet shortcrust pastry	5. Line a 9 in diameter foil plate with 8 oz (250 g).
		6. Add 1 lb 2 oz (550 g) fruit filling.
		7. Top with 7 oz (200 g) pastry.
½ pint (250 cc)	Egg wash	8. Brush with egg wash.
	Sugar to dredge	9. Sprinkle with sugar.
		10. Wrap in suitably labelled foil.
		11. Put into freezer for 90 minutes.
		12. Store in holding fridge until required.

Lid Notes
Description: Apple and Blackcurrant Pie
No. of portions: 8
Oven setting: 5
Oven time: 35 minutes
Lid off
Pack weight: 1 kg

APRICOT CRUMBLE (560 PORTIONS)

16 × A10	Apricots	1. Drain the apricots. Make the juice up to 16 pints (9 litres) with water if necessary.
12 lb (5·5 kg)	Sugar	2. Add the sugar to the juice and bring to the boil.
1½ lb (675 g)	Freeze 'n' flow starch Cold water to mix	3. Mix the freeze 'n' flow to a smooth paste with cold water. Stir into the boiling syrup; bring back to the boil, stirring all the time. Reduce the heat and simmer until the mixture clears.
		4. Pour the sauce on to the apricots and blend them together.
45 lb (21 kg)	Crumble mix*	5. See the recipe for crumble mix and make up as instructed.
		6. Weigh 1 lb 6 oz (600 g) of apricot filling into each 9½ in round foil container.
		7. Top with 1 lb 2 oz (500 g) crumble mix.
		8. Lid and seal.
		9. Put into freezer for 90 minutes.
		10. Store in holding fridge until required.

* 45 lb of flour, i.e., 3 × recipe quantities.

Lid Notes
Description: Apricot Crumble
No. of portions: 8
Oven setting: 5
Oven time: 35–40 minutes
Lid off
Pack weight: 1100 g

APRICOT TART (320 PORTIONS)

8 × A10 (3 kg)	Apricots	1. Drain the apricots. Make the quantity of juice up to 8 pints (4·5 litres) with water if necessary.
6 lb (2·75 kg)	Sugar	2. Put the sugar into the syrup and bring to the boil.
12 oz (350 g)	Freeze 'n' flow starch Cold water	3. Mix the freeze 'n' flow to a smooth paste with cold water. Stir into the syrup. Bring back to the boil stirring all the time. Reduce the heat and simmer until the mixture clears.
		4. Add the thickened syrup to the apricots and stir until evenly distributed.
20 lb (10 kg)	Sweet shortcrust pastry	5. Line 9 in diameter pie foils with 8 oz (250 g) pastry.
		6. Weigh in 1 lb 2 oz (550 g) pie filling.
		7. Top with 7 oz (200 g) pastry.
½ pint (250 cc) 8 oz (250 g)	Egg wash Sugar	8. Brush with egg wash and dredge with sugar.
		9. Wrap with foil and label.
		10. Place in freezer for 90 minutes.
		11. Store in holding fridge until required.

Lid Notes
Description: Apricot Tart
No. of portions: 8
Oven setting: 5
Oven time: 35 minutes
Lid off
Pack weight: 1 kg

BAKED PLUM SPONGE (312 PORTIONS)

8 × 3 kg	Red plum halves*
4 lb (1·8 kg)	Sugar
	Cold water

1. Drain the plums and put the juice into a saucepan. Add the sugar and bring to the boil. Make up to 8 pints (4·5 litres) with water.

| 12 oz (350 g) | Freeze 'n' flow starch |

2. Reconstitute the freeze 'n' flow with cold water and stir into the boiling syrup. Bring to the boil, reduce the heat and simmer until the starch is cooked.

3. Pour the thickened sauce on to the drained plums and blend them together. Allow to cool.

| 28 lb (12·5 kg) | Sponge mix |
| 11¼ pints (6·4 litres) | Water |

4. Reconstitute the sponge mix following manufacturers instructions.

5. Weigh 1 lb 2 oz (500 g) of plum mixture into each 9½ in round container.

6. Top the plums with 1 lb 2 oz (500 g) of sponge mixture.

7. Smooth and level the top of the sponge with a wet scraper.

8. Lid and seal.

9. Put into freezer for 90 minutes.

10. Store in holding fridge until required.

* Use imported Red Plum Halves if available. These are stoneless and firmer than English varieties.

Lid Notes
Description: Baked Plum Sponge
No. of portions: 9
Oven setting: 5
Oven time: 35 minutes
Lid off
Pack weight: 1 kg

BAKEWELL TART (448 PORTIONS)

28 lb (12·5 kg)	Sponge mix
11¼ pints (6·4 litres)	Water
1 oz (25 g)	Almond essence
7 lb (3·5 kg)	Mixed fruit jam
28 lb (12·5 kg)	Sweet shortcrust pastry

1. Mix the sponge mix following manufacturers instructions and add the almond essence.

2. Heat the jam to a pouring consistency.

3. Line the 9 in diameter foil plates with 8 oz (250 g) pastry.

4. Add 2 oz (50 g) jam and cover the bottom of the pastry.

5. Add one 15 fl. oz ladle of sponge mix (12 oz (340 g) in weight). Smooth with a wet scraper.

6. Wrap with suitably labelled foil sheets.

7. Put into freezer for 90 minutes.

8. Put into holding fridge until required.

Lid Notes
Description: Bakewell Tart
No. of portions: 8
Oven setting: 4
Oven time: 35 minutes
Lid off
Pack weight: 640 g

BREAD AND BUTTER PUDDING (320 PORTIONS)

8	Thin sliced loaves	1. Melt the margarine and quickly dip the slices of bread into it. Don't let the bread get sodden.
4 lb (1·75 kg)	Margarine	
5 lb (2.25 kg)	Sultanas	2. Cut the slices of bread in half. Arrange 5 × $\frac{1}{2}$ slices on the bottom of each $9\frac{1}{2}$ in sq shallow container, sprinkle on 1 oz (25 g) sultanas. Arrange another 5 × $\frac{1}{2}$ slices of bread on the top and sprinkle with another 1 oz (25 g) sultanas.
120	Eggs	3. Crack the eggs into a large mixing bowl, add the sugar and beat together.
$4\frac{1}{2}$ lb (2 kg)	Sugar	
6 gals. (28 litres)	Milk	4. Add the milk to the eggs and sugar. Mix to form a custard.
	Mixed spice	5. Pour 2 × 15 oz (850 cc) ladles of custard on to each foil container. Sprinkle with mixed spice.

6. Lid and seal.

7. Put into freezer for 90 minutes.

8. Store in holding fridge until required.

Lid Notes
Description: Bread and Butter Pudding
No. of portions: 8
Oven setting: 4
Oven time: 40 minutes
Lid off
Pack weight: 1 kg

CRUMBLE MIXTURE

7½ lb (3·5 kg) Margarine
5 lb (2·3 kg) Sugar

15 lb (7 kg) Plain flour
1 tablespoon Salt
12 oz (350 g) Raising powder

1. Put the margarine and sugar into a large mixing bowl. Beat until white and creamy.

2. Sieve the flour, salt and raising powder into the creamed margarine and sugar.

3. With the machine on its slowest speed carefully mix the ingredients together until they resemble breadcrumbs. Turn the machine off immediately. *Do not overmix.*

4. Put the mixture into a refrigerator to cool before use, this way it will handle better.

Note: This mixture represents 15 lb (7 kg) of crumble mixture on the recipes.

DUTCH APPLE TART (320 PORTIONS)

6 × A10	Solid pack apples	1. Slice the apples with the slicing attachment into a suitably sized bowl.
6 pints (3·5 litres) 6 lb (2·75 kg) 2 oz (50 g) 9 oz (255 g)	Water Sugar Mixed spice Freeze 'n' flow starch	2. Put the water, sugar and spice into a pan, bring to the boil and thicken with the reconstituted freeze 'n' flow.
3 lb (1·35 kg) 3 lb (1·35 kg)	Sultanas Currants	3. Mix the sultanas and currants into the sauce and stir into the sliced apples. Blend together evenly and allow to cool.
20 lb (10 kg)	Sweet shortcrust pastry	4. Line the 9 in diameter foil plates with 8 oz (250 g) pastry.
		5. Add 1 lb 2 oz (500 g) apple pie filling.
		6. Cover with 7 oz (200 g) pastry.
½ pint (250 cc)	Egg wash	7. Brush with egg wash.
	Sugar to dredge	8. Sprinkle with sugar.
		9. Wrap with suitably labelled foil.
		10. Put into freezer for 90 minutes.
		11. Store in holding fridge until required.

Lid Notes
Description: Dutch Apple Tart
No. of portions: 8
Oven setting: 5
Oven time: 35 minutes
Lid off
Pack weight: 1 kg

PEAR UPSIDE DOWN (312 PORTIONS)

6 × A10	Pears	1. Drain and slice the pears.
1½ lb (700 g)	Melted margarine	2. Brush the 9½ in round foil containers liberally with margarine.
4 lb (1·8 kg)	Demerara sugar	3. Coat the bottom of the foils with 2 oz (50 g) demerara sugar.
		4. Cover the sugar with 8 oz (225 g) sliced pears.
28 lb (12·5 kg)	Sponge mix	5. Blend the sponge mix and water following the manufacturers instructions.
11¼ pints (6·4 litres)	Water	
		6. Weigh 1 lb 2 oz (500 g) sponge mix into each container. Smooth and level with a wet scraper.
		7. Lid and seal.
		8. Freeze for 90 minutes.
		9. Store in holding fridge until required.

Note: This sweet should be turned out on to a plate after cooking.

Lid Notes
Description: Pear Upside Down
No. of portions: 9
Oven setting: 5
Oven time: 40 minutes
Lid off
Pack weight: 775 g

PINEAPPLE UPSIDE DOWN (312 PORTIONS)

1½ lb (700 g)	Melted margarine	1. Brush the 9½ in round foil containers liberally with margarine.
4 lb (1·8 kg)	Demerara sugar	2. Coat the bottom of the foils with 2 oz (50 g) demerara sugar.
4 × A10 (112–120 count)	Pineapple rings	3. Drain the pineapple rings and arrange 10 rings in each container.
8 oz (225 g)	Glacé cherries	4. Put half a glacé cherry into the centre of each ring.
28 lb (12·5 kg) 11¼ pints (6·4 litres)	Sponge mix Water	5. Blend the sponge mix and water following the manufacturers instructions.
		6. Weigh 1 lb 2 oz (500 g) of sponge into each container. Smooth with a wet scraper.
		7. Lid and seal.
		8. Freeze for 90 minutes.
		9. Store in holding fridge until required.

Note: This sweet should be turned out on to a plate after cooking.

Lid Notes
Description: Pineapple Upside Down
No. of portions: 9
Oven setting: 5
Oven time: 40 minutes
Lid off
Pack weight: 775 g

RHUBARB AND GINGER TART (320 PORTIONS)

6 pints (3·5 litres) 6 lb (2·75 kg) 2 oz (50 g)	Water Sugar Ground ginger	1. Bring the water, sugar and ginger to the boil in a saucepan.
9 oz (255 g)	Freeze 'n' flow starch	2. Reconstitute the freeze 'n' flow with cold water and stir into the syrup. Bring back to the boil, reduce the heat and simmer until the mixture clears.
6 × A10	Solid pack rhubarb	3. Open the rhubarb and put into a suitably sized bowl. Add the thickened syrup and stir well. Allow to cool.
20 lb (10 kg)	Sweet shortcrust pastry	4. Line the 9 in diameter pie foils with 8 oz (250 g) pastry.
		5. Add 1 lb 2 oz (550 g) rhubarb pie filling.
		6. Cover with 7 oz (200 g) pastry.
$\frac{1}{2}$ pint (250 cc)	Egg wash	7. Brush with egg wash.
	Sugar to dredge	8. Sprinkle with sugar.
		9. Wrap with suitably labelled foil sheets.
		10. Put into freezer for 90 minutes.
		11. Store in holding fridge until required.

Lid Notes
Description: Rhubarb and Ginger Tart
No. of portions: 8
Oven setting: 5
Oven time: 35 minutes
Lid off
Pack weight: 1 kg

RICE PUDDING (132 PORTIONS)

6 gals. (28 litres)	Milk
3 lb (1·5 kg)	Sugar
5 lb (2·5 kg)	Rice
12 oz (350 g)	Margarine
1 tablespoon	Salt

1. Pre-heat a suitably sized double boiler.

2. Put all the ingredients into the boiler, stir well. Bring to the boil, stirring regularly to stop the rice from congealing. Reduce the heat to a simmer and continue cooking, stirring occasionally, until the rice is soft.

3. Check for flavour and consistency.

4. Weigh 3 lb (1·35 kg) of rice pudding into each 9½ in sq shallow container.

5. Lid and seal.

6. Put into freezer for 90 minutes.

7. Store in holding fridge until required.

Lid Notes
Description: Rice Pudding
No. of portions: 6
Oven setting: 5
Oven time: 40 minutes
Lid on
Pack weight: 1·35 kg

SAGO PUDDING (132 PORTIONS)

6 gals. (28 litres)	Milk
3 lb (1·5 kg)	Sugar
5 lb (2·5 kg)	Sago
12 oz (350 g)	Margarine
1 tablespoon	Salt
1 teaspoon	Vanilla essence

1. Pre-heat a suitably sized double boiler.
2. Put all the ingredients into the boiler. Stir well. Bring to the boil stirring regularly to stop the sago from congealing. Reduce the heat to a simmer and continue cooking, stirring occasionally, until the sago is soft.
3. Check for flavour and consistency.
4. Weigh 3 lb (1·35 kg) of sago pudding into each $9\frac{1}{2}$ in sq shallow container.
5. Lid and seal.
6. Put into freezer for 90 minutes.
7. Store in holding fridge until required.

Lid Notes
Description: Sago Pudding
No. of portions: 6
Oven setting: 5
Oven time: 40 minutes
Lid on
Pack weight: 1·35 kg

STEAMED MARMALADE SPONGE (312 PORTIONS)

28 lb (12·5 kg)   Sponge mix
11¼ pints (6·4 litres)   Water

17 lb (7·5 kg)   Marmalade

1. Put the sponge mix into a mixing bowl and blend with the water following manufacturers instructions.

2. Warm the marmalade until it pours. Do not burn.

3. Weigh 7 oz (175 g) of marmalade into each 9½ in diameter deep round container.

4. Add 1 lb 4 oz (550 g) sponge mix.

5. Put into freezer for 90 minutes.

Lid Notes
Description: Steamed Marmalade Sponge
Cooking time: 2 hours
Lid on
Pack weight: 825 g

STEAMED SYRUP SPONGE (312 PORTIONS)

28 lb (12·5 kg)   Sponge mix
11¼ pints (6·4 litres) Water

17 lb (7·5 kg)   Syrup

1. Put the sponge mix into a mixing bowl and blend with the water following manufacturers instructions.

2. Warm the syrup in the bainmarie.

3. Put 7 oz (175 g) syrup into each 9½ in diameter deep container.

4. Add 1 lb 4 oz (550 g) sponge mixture.

5. Lid and seal.

6. Blast freeze for 90 minutes.

Lid Notes
Description: Steamed Syrup Sponge
Cooking time: 2 hours
Lid on
Pack weight: 825 g

SYRUP TART (560 PORTIONS)

44 lb (20 kg)	Syrup	1. Warm the syrup in the bainmarie until it is thin and pours easily.
10	Lemons	2. Grate the rind off the lemons and extract the juice.
16 lb (7·25 kg)	White breadcrumbs	3. Mix together the breadcrumbs, lemon juice and rind and syrup in a mixing bowl.
35 lb (17·5 kg)	Sweet shortcrust pastry	4. Line the 9 in diameter pie plates with 8 oz (250 g) pastry.
		5. Add 13 oz (400 g) syrup filling.
		6. Wrap in suitably labelled foil sheets.
		7. Freeze for 90 minutes.
		8. Put into holding fridge until required.

Lid Notes
Description: Syrup Tart
No. of portions: 8
Oven setting: 4
Oven time: 35 minutes
Lid off
Pack weight: 650 g

SWEET SHORTCRUST PASTRY

7 lb (3·5 kg)	Margarine
3 lb (1·5 kg)	Lard
4 lb (2 kg)	Sugar
2 pints (1 litre)	Cold water
1 tablespoon	Salt
20 lb (10 kg)	Plain flour
10 oz (312·5 g)	Golden raising powder
2–2½ pints (1·5 litres)	Water

1. Cream the margarine, lard, sugar and salt in the mixer. Scrape down and gradually beat in the water. Continue beating until the fat is white and fluffy.

2. Sieve the flour and raising powder into the creamed lard. Put the mixer onto its slowest speed and blend the ingredients together until they resemble bread-crumbs.

3. Carefully bind the ingredients together with the water to form a short pastry. This takes about 45 seconds on slow speed. Do not overmix or the pastry will tend to shrink and harden when cooked.

4. Allow to stand for 20 minutes before using.

Note: This mixture represents 20 lb (10 kg) sweet shortcrust pastry on the recipes.

TAPIOCA PUDDING (132 PORTIONS)

6 gals. (28 litres)	Milk
3 lb (1·5 kg)	Sugar
5 lb (2·5 kg)	Tapioca
12 oz (350 g)	Margarine
1 tablespoon	Salt
1 teaspoon	Vanilla essence

1. Pre-heat a suitably sized double boiler.

2. Put all the ingredients into the boiler, stir well. Bring to the boil stirring regularly to stop the tapioca from congealing. Reduce the heat to a simmer and continue cooking, stirring occasionally, until the tapioca is soft.

3. Check for flavour and consistency.

4. Weigh 3 lb (1·35 kg) of tapioca pudding into each 9½ in sq shallow container.

5. Lid and seal.

6. Put into freezer for 90 minutes.

7. Store in holding fridge until required.

Lid Notes
Description: Tapioca Pudding
No. of portions: 6
Oven setting: 5
Oven time: 40 minutes
Lid on
Pack weight: 1·35 kg

PART III

Catering Costing System

INTRODUCTORY REMARKS

One of the most serious problems facing large-scale caterers is the almost total lack of financial information in time to take the necessary management decisions. Although the costing and accounting systems at Keele have always been tight and effective, the onset of continuing inflation made management realise that these systems were simply not fast enough to allow adjustments to be made to, for example, buying and production that would have the maximum financial effect. It was therefore decided to devise a system of costing that utilised a small part of the University's computing resources. Although the formal objectives of this exercise are listed below it should be emphasised that the firm intention was always to ensure that accurate and as up to date as possible information be provided without creating additional and unnecessary paperwork. In order to achieve this the existing systems were carefully examined and were found to be generally satisfactory, and as a result few fundamental changes were made. Although new forms were designed, procedures remained essentially the same and as a result, the system was introduced with little fuss and no confusion.

Today, the catering managers can look at their weekly print outs and make decisions based upon facts and not assumptions. The result has been a marked increase in effectiveness throughout the organisation.

OBJECTIVES OF THE SYSTEM

To supply management with information regarding:

1. Types of food in store, and quantities and values for each type.
2. Current commodity cost information by reference to the most recent prices paid.
3. Current food cost of a range of dishes.
4. Current food cost of each daily menu.
5. Cost and type of food usage by refectories.
6. Verification of purchases of food by reference to the creditors payment system.
7. Appropriate information similar to the above in relation to alcoholic drinks (i.e., spirits, wines and ales).

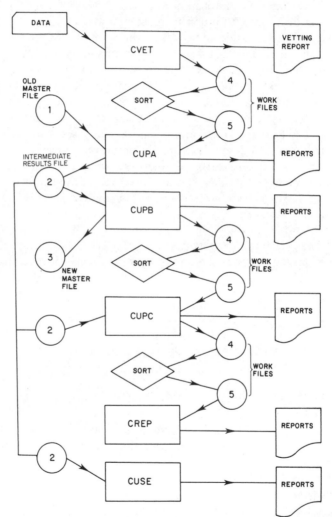

Fig. 7.   Order of Data Processing.

THE COMPUTER SYSTEM

In designing a computer system for this application, it was necessary to work within the framework of the computer hardware, software and processing time available at the University Computer Centre. This dictated the design of the system round a magnetic tape based serially processed master file.

Subsequently, owing to a change of computer equipment, the system has been converted to use disc based serial files and is now run on an ICL 2903 computer located in the Administration Data Processing Unit.

The system is run weekly and comprises a suite of six Cobol programs and three sort programs running under the ICL operating system. In order to maintain processing continuity it is essential to run the system every week. In addition, special programs are used from time to time in connection with stocktaking and audit functions.

Data for input to the system are entered by the Catering Department Staff on a series of specially designed forms. Batches of complete forms are submitted to the Data Processing Unit at weekly intervals when they are prepared for computer processing by key-punch operators using ICL Direct Data Entry equipment or alternatively, card-punch equipment.

Processing of the data by the suite of programs is in the order shown in Fig. 7 and reports are printed at appropriate stages. Each report is identified at the top left of each page by the current date, week and year number and the name of the program producing the report. The year number is the last two digits of the calendar year in which the current financial year ends.

The input documents and the computer reports produced are described in detail in the following pages.

INPUT DOCUMENTS

The various forms used to input data to the computer system also serve to record the receipt of goods into store, the production of dishes by the central production unit and the issue of commodities to the various user locations. They are essentially similar to the documents which were formerly used to control the usage of food under the inefficient manually operated system which preceded the present computer based system. The introduction of computer processing has therefore caused no significant extra paperwork for the catering staff.

Care was taken to design forms which kitchen and storekeeping staff could complete without undue difficulty, but which were also suitable for rapid and unambiguous transcription by the data preparation staff.

During the implementation stage there were naturally some delays in the preparation of the input documents although the general level of accuracy was always good. However, once the causes of delay were identified, the catering staff soon found that the computer system gave reliable information much more quickly than the previous manual system and delays in the completion of input documentations were reduced to a minimum whilst maintaining an excellent level of accuracy.

The notes which follow are taken from instructions prepared when the system was first introduced as guidance to the staff responsible for its implementation. The numbers given under 'Detail' refer to the column numbers printed on the sample forms at the head of each column.

R01 Goods Received (Fig. 8)

Completed by Storekeeper.

Function To record goods received for catering, detailing type, quantity and cost for each item.

Detail 2-part pad of pre-numbered forms. The white copy sent to Finance Office, the green copy retained in catering stores and filed to form a 'Goods Received Book'.

1–3 Pre-printed form number.

4 Leave blank.

5–8 Use the first 4 letters of suppliers name, (e.g., Titley—TITL). Alongside 'From:-' may be shown supplier's full name and any detail such as invoice number, etc.

9–13 Use 'Order Number' taken from the University Order relevant to the invoice(s)—use numbers only—i.e., for order number R8864, show 08864.

Use separate R01 for each order number quoted on invoice coding slip; the same R01 may be used for several invoices with a common order number.

14–17 Pre-printed sheet number.

18–23 Date of receipt of goods—day, month, year—show 29 07 76; 06 12 76.

24–28 Stock number for the item as shown in the 'Stores Ledger Listing' (q.v.)—baked beans, if A10 size, show 65779; if 7·75 oz size, show 65747.

29–34	The quantity of the item received, corresponding with the description of package shown in the stores ledger listing for baked beans 65779; the unit is *one* A10 tin, so if 5 cases of 6 A10 tins are received, the number to be shown as received is 30.
35–40	Cost is the full cost (excluding any V.A.T.) of the item purchased.
41	Only complete this box when any item *previously recorded as received* is returned to the supplier. Use 'R'.
42–44	Complete only if the item has been delivered by the supplier direct to a kitchen, and has never been put into central store.

The last line on the form should be completed as follows:

Stock number 99994 is programmed to indicate that the following numbers are *totals* for the sheet.

Show total of quantities received and of cost (columns 29–40).

If the totals are debits, show 'R' in column 41.

If the totals are credits, leave column 41 blank.

R01X Daily Catering Production Return (Fig. 9)

Completed by	Production Manager/Unit Storekeeper.
Function	To record the daily production of items by the Production Unit for transfer to the main holding fridges.
Detail	Production code: 06126. Day: Monday.
18–23	Date of production: 06 12 76.
Details	For catering department use. Description of items produced.
24–28	Stock number for Apple Tart as shown in Production Unit section of stores ledger listing (q.v.): 98239.
29–34	Total number of packs put into store. If any items go direct from Production Unit to a kitchen to satisfy an emergency need, those items must be included here as production and requisitioned out to the kitchen using Form R02 (q.v.)
41	If any items are returned from the main holding fridge to the Production Unit as being unfit for storage, then complete 29–34 as above and show 'R' in column 6.

Note: Any returns to main holding fridge from outside kitchens will be shown on R02, not this form R01X.

The final entry on this form, with stock number 99994 is the total number of packs recorded on the sheet.

R02 Catering Stores Requisition (Fig. 10)

Completed by Kitchen Supervisors, Storekeepers.

Function   Request from kitchens and other places for goods both from dry goods store, and production unit refrigerated store.

Detail   2-part pad of pre-numbered forms. Both copies sent from kitchens to stores: white copy sent to Finance Office, pink copy returned to kitchen with goods supplied, then acting as a delivery note.

1–3   Pre-printed number.

4   Leave blank.

5–7   Pre-printed with the name of the kitchen.

8–11   Pre-printed sheet number.

12–17   Date of supplying goods, or the first day of supply where items are despatched over a period of time, not exceeding a week. Day, month and year—show 29 07 76.

Description   For catering department use—show a short, but clear description of the quantity and type of items ordered, e.g. 3 × 6 A10 Beetroot.

18–22   Stock number for the items as shown in the 'stores ledger listing'.

23–28   The quantity of the item supplied, corresponding with the description of package as shown in the 'stores ledger listing'.

29   Only complete this box when any item *previously recorded* as requisitioned is returned to the central stores, or the main holding fridge—use 'R'.

30–32   Complete only if the item is taken from one kitchen and sent to another place without going via stores.

When completed, appropriate signatures and dates must be shown in the spaces provided at the foot of the form.

R03 Menu Plan (Fig. 11)

Completed by Production Manager.

Function   To detail the dishes to be offered at each meal, and show the likely uptake of each dish. To create a menu data record.

Detail

1–3	Pre-printed form number.
4	Leave blank.
5–7	Menu number thus:

First number is the week number of the menu cycles—
    term time cycle: 1–5
    vacation cycle: 7–8
Second number is the day of the week—
    1 Monday, 2 Tuesday, 7 Sunday.
Third number is the meal number—
    1 Breakfast
    2 Morning coffee
    3 Luncheon
    4 Afternoon tea
    5 Dinner.

12–17	Leave blank.
18–19	Pre-printed item number.
20–24	Dish number as shown on 'dish cost summary'.
25–30	Show percentage likely uptake of each dish for each kitchen. Total uptake in any item section (e.g. starter, sweet, etc.) must not exceed 99, but it may be less than 99. Proportions of less than 2 % are not permitted.

R04 Recipe Coding Sheet (Fig. 12)

Completed by Production Manager.

Function   To detail recipe ingredients showing type and quantity. To create a recipe data record. For Production Unit items, to create a stock ledger record, thereby functioning as an R05 stock coding sheet (q.v.).

Detail   Because of the large amount of information needed to create a full recipe data record, the one form is in two parts, with boxes (1) and (2) common to both parts.

1–3	Pre-printed form number.
4	Leave blank.
5–9	Recipe number. Dish recipes codes start with a number zero (0). Production Unit production recipes start with a number nine (9).

Part 1

10–11	Pre-printed with zeros.
12–14	The number of dish portions the recipe will serve.
15–38	Recipe name for Production Unit item of production.
or	
15–46	Recipe name for presented dishes.
47–49	For Production Unit recipes only; the number of portions to be put into each pack. For presented dishes, leave blank.
50–58	Leave blank.
59–64	For Production Unit recipes only; the food cost of one pack of prepared food in £...p...
65–66	For Production Unit recipes only; classification by food group.

Part 2

Ingredients	Short verbal description.
10–11	Pre-printed item number, 20 items per form. For any recipe with more than 20 items, continue with another form, completing columns 1 to 9, then continuing with item 01.
12–16	Stock number for the item as shown on the 'stores ledger listing'.
17	Unit of quantity to follow:
	1 for kilograms and grams
	2 for litres and ml
	3 for lbs and oz
	4 for pints and fl. oz
	5 for numbers
18–23	Quantity of the item used.

R05 Stock Coding Sheet (Fig. 13)

Completed by	Production Manager.
Function	To detail each separate item of stock used within the catering system. To create a stock ledger record.
Detail	For economy of paper 3 stock coding sheets can be printed on one page.
1–3	Pre-printed form number.
4	Leave blank.

5–9	Stock number—selected from a pre-printed list of self-checking numbers kept in the Finance Department. The number for any particular item of stock is selected from the ranges as shown:

    1 direct issues
    2 frozen foods
    3 meat

    $\left.\begin{matrix}4\\5\\6\end{matrix}\right\}$ dry goods

    $\left.\begin{matrix}7\\8\end{matrix}\right\}$ beer, spirits, wines, tobacco

10–14	Leave blank.
15–38	Description of the item.
39	Leave blank.
40–46	Package size of each unit of the item, e.g., A10; 3 kg; Box 24.
47	Unit of package:

    1 for kilograms and grams
    2 for litres and ml
    3 for lbs and oz
    4 for pints and fl. oz
    5 for numbers.

48–53	Quantity in each unit of the item, e.g., A10 tin of fruit 6 lb 10 oz; 3 kg; 24.
54–56	Receipt multiple—show *minimum* number of units of delivery for the item—e.g. 32 kg flour, minimum receipt multiple is 1; 1 case of baked beans, the minimum receipt multiple is 6.
57	The number shown indicates the types of issues and receipts allowable for the item:

    1 direct issues not permitted
    2 direct issues permitted
    3 issues only
    4 receipts only
    blank = 1.

58	The number shown indicates the pricing system for the issues:

    1 issues made at average price
    2 issues made at standard price
    blank = 1.

59–64   Standard issue price for use when:
               (a)   creating a record for a new item;
               (b)   setting the standard price for items issued at that
                     standard price.

               To a maximum of £999 99·9.
65–66   Class of food to which the item is to be allocated, e.g., baked
               beans to class 34 vegetables—canned.

Invoice Coding (Fig. 14)

Completed by Production Manager, Storekeeper, Clerk.
Function   To authorise payment of the attached invoice by Finance
               Office and debit accounts as shown by code numbers.
Detail   When goods are ordered from a supplier, details are shown
               on a pre-numbered order form.
                 Storekeeper checks invoice against delivery and authorises
               payment by writing order number on invoice.
                 Clerk codes invoice coding slip thus:
                 Main code—use 72 for all food items.
                 Departmental code⎫
                 Sub code   ⎬—use numerics of the order number.
                 Detail code   ⎭
                 e.g. Order number—8664, code—72 00 86 64;
                     Order number—19528, code—72 01 95 28.
                 Code any non-food items (e.g. V.A.T.) according to
               current University instructions.
                 Code 'Drinks' according to current University
               instructions.

UNIVERSITY OF KEELE

CATERING

GOODS RECEIVED

From: -

FORM NUMBER	SUPPLIED BY	ORDER NUMBER	SHEET NUMBER	DATE RECEIVED
1 14	6	9	14	18 23
R01	T I T L 0.8.8.6.4		1234	0.1 0.3 7.9

DETAILS	STOCK NUMBER	QUANTITY RECEIVED	COST (EXCLUDING VAT)	RETURN	DIRECT TO
	24	29	35 40	42	44
TOMATOES	1.9.6.4.7	. . . 1 3	. . 3 0.0		L I N
	. . . .	. . . .	. . .		. . .
	. . . .	. . . .	. . .		. . .
	. . . .	. . . .	. . .		. . .
	. . . .	. . . .	. . .		. . .
	. . . .	. . . .	. . .		. . .
	9 9 9 9 4	. . . 1 3	. . 3 0.0 R		

Storekeeper . . . .

FIG. 8.   R01 Goods Received.

UNIVERSITY OF KEELE

DAILY CATERING PRODUCTION RETURN

Production Code 06126   Day MONDAY

FORM NUMBER	IDENT	DATE PRODUCED
1 4	9 13	18 23
R01X	99999	06 12 76

DETAILS	No. of Packs Achieved	Portions per Pack	STOCK NUMBER	PACKS INTO STORE	R
			24	29 34	41
APPLE TART	28	8	9.8.2.3.9	. . . 2 8	
			. . . .	. . . .	
			. . . .	. . . .	
			. . . .	. . . .	
			9 9 9 9 4	. . . 2 8 R	

Production Signature -

Storekeeper -

FIG. 9. R01X Daily Catering Production Return.

UNIVERSITY OF KEELE

CATERING STORES REQUISITION

Supply to:-

LINDSAY REFECTORY

FORM NUMBER	SUPPLIED TO	REQUISITION NUMBER	DATE SUPPLIED
1 4	5	8	12 17
RO2	LIN	1200	29 07 76

REQUISITIONED		DESCRIPTION	STOCK NUMBER	QUANTITY SUPPLIED	RETURN	FROM	
QTY.	UNIT		18	23 28		30	32
3 X 6	A10	BEETROOT	6 6 0 1 0	1 8			

Goods Ordered by ....

Date Ordered ....   Goods Received by ....

Storekeeper.......................... Date Received ....

| 9 9 9 9 4 | 1 8 R |

FIG. 10.   R02 Catering Stores Requisition.

UNIVERSITY OF KEELE	FORM NUMBER	MENU NUMBER		CODES				SPECIFIC DATE		
CATERING DEPARTMENT	1 4	5 7	A	B	C	D	12			17
MENU PLAN	R03	1 1 3					2 9 0 7 7 6			

COURSE	DISH	ITEM 18	DISH NUMBER 20	STANDARD MIX HOR 25	LIN 27	HAW 29
STARTER		11				
		12				
		13				
MAIN	ASSORTED COLD MEATS	21	0 4 8 1 6	4 0	4 0	4 0
	STEAK AND MUSHROOM PIE	22	0 3 9 5 7	1 4	1 4	1 4
	FILLET OF FISH BONNE FEMME	23	0 2 5 9 0	6	6	6
	MACARONI AU GRATIN	24	0 4 3 2 6	3	2	3
	ASSORTED COLD SALAD CHEESE	25	0 5 9 9 3	6	6	6
	BREADED VEAL CUTLETS	26	0 3 1 5 3	1 3	1 2	1 3
	CHICKEN AND MUSHROOM CASSEROLE	27	0 3 2 5 9	1 6	1 7	1 6
	VEGETABLE CUTLETS	28	0 5 7 4 6	1	2	1
VEGETABLE	SALAD	31	0 6 5 1 7	4 6	4 6	4 6
	FRENCH BEANS	32	0 6 0 4 1	3 0	2 9	2 9
	BUTTERED PARSNIPS	33	0 6 2 3 5	2 3	2 4	2 4
		34				
POTATO	POTATO SALAD	41	0 6 8 9 1	4 6	4 6	4 6
	PARSLEY POTATOES	42	0 6 8 1 3	3 4	3 5	3 4
	DUCHESSE POTATOES	43	0 6 7 3 2	1 9	1 8	1 9
		44				
SWEET	SEMOLINA PUDDING	51	0 8 3 8 4	7	6	8
	FRESH FRUIT	52	0 7 5 1 0	4 4	4 4	4 4
	MARMALADE SPONGE AND CUSTARD	53	0 7 7 7 5	4 8	4 9	4 7
		54				
		55				
CHEESE	CHEESE AND BISCUITS	61	0 9 5 1 8	9 0	9 0	9 0
		62				
COFFEE	COFFEE AND TEA	71	0 9 0 1 0	9	9	9
		72				

FIG. 11.   R03 Menu Plan.

UNIVERSITY OF KEELE	FORM NUMBER	RECIPE NUMBER
CATERING DEPARTMENT	1 4	5 9
RECIPE CODING SHEET	R04	0 3 0 2 6

FIRST CARD

ITEM 10	TO SERVE 12		NAME OF RECIPE 15 39	46
00	8		BEEF HØT PØT	

PORTIONS PER PACK 47	LABOUR UNITS 50	COOKING UNITS 53	FREEZING UNITS 56	STANDARD PACK COST 59	CLASS 65	
8						

INGREDIENTS	ITEM 10	STOCK NUMBER 12	UNIT 17	QUANTITY 18 23
BEEF HØT PØT	01	9 2 1 2 4	5	. 1 .
CRESS	02	1 9 3 0 1	1	. . 8
	03			
	04			
	18			
	19			
	20			

Checked **by**

....................................Catering Officer

FIG. 12.   R04 Recipe Coding Sheet.

FORM NUMBER	STOCK NUMBER
1 4	5 9
R05	2 8 7 9 7

ITEM 10	CODES A B C		DESCRIPTION 15 39	S	PACKAGE 40 46
00			CØRNISH PASTIES		4 0

UNIT 47	QUANTITY IN PACKAGE 48	RECEIPT MULTIPLE 54	R 57	P 58	STANDARD ISSUE PRICE 59	CLASS 65	
5	4 0	1	2	2	. 3 1 2 0	0 7	

FIG. 13.   R05 Stock Coding Sheet.

FOR USE BY FINANCE OFFICE			FOR USE BY DEPARTMENT					
Internal Audit	Invoice	Coding	Passed for Payment				Price	
			Dept. Check	Qty.	Qual.		£	p
CREDITOR'S NAME			Order Nos.		CODES			
Period	Creditor's Number							
C 5 1								
Finance Reference								
Invoice Details								
Departmental Refnc's.			INVOICE TOTAL					

INVOICE CODING SLIP

NRW/SJP (18.11.70)

FIG. 14. Invoice Coding Slip which is attached to a complete invoice.

COMPUTER REPORTS

A very wide range of reports is produced by the computer unit to meet the complex requirements of the catering service.

Many of the regular reports are concerned with the routine but vitally important functions of ensuring accounting accuracy and maintaining efficient stock control whilst others are aids to purchasing policy, menu planning, budget control and catering management.

Several hundred sheets are printed each week in a computer run which takes about one and a half hours. The system is kept under constant review and the scope and content of reports are varied from time to time as specific needs become apparent.

In the following pages a reasonably comprehensive overview of the range of reports currently produced is presented but it is not practicable to include examples of all the items reported on.

Pre-sort Vetting (Fig. 15)

This mundane document plays a vital part in ensuring accuracy of data recording and transcription. It is used primarily by the data processing staff to identify and, following investigation, to rectify errors in the raw data before the data are sorted into the order required for processing by the main programs in the suite.

An important feature of the system is the use of self-checking numbers (numbers which incorporate a check digit), to enable detection of incorrect stock numbers and recipe number. This system gives 100% detection of single transcription and single transposition errors and about 90% detection of random errors in the specifying of stock numbers. The vetting program also verifies sheet totals and embodies many other checks on the accuracy of the data. Some error requests result in the rejection of the line of data, others are treated as warnings.

A typical weekly run contains about a thousand lines of data and each line in a batch of data is serially numbered to facilitate the rapid correction of errors by the Direct Data Entry Keystation Operator.

The pre-sort vetting report forms a valuable link in the audit trail and is held in the Catering Costs Office together with the input documentation and certain other reports until after the completion of the annual audit.

Transaction Listing (Figs. 16 and 17)

This report lists in detail each commodity for which a transaction has occurred. It is at this stage that very comprehensive vetting of receipt values and prices takes place. The system will reject an attempt to up-date the file with receipts at abnormally high or low prices thus avoiding major errors. Routine rise and fall of prices is catered for and these are indicated.

The computer calculates the value of all issues of food from stock and also the value of items produced by the production unit. In the event of issues exceeding receipts (because of delays in processing of receipt documentation) issues will be made at the latest average price. Subsequently the computer will, if necessary, calculate a price adjustment. The weekly total of such adjustments is apportioned between the major cost centres in the same ratio as the total food issues to date.

Where issues exceed receipts causing the stock balance to go negative, the computer prints a warning message.

The computer records three types of prices, average price (marked A), cost price (marked C) and standard price (marked S). Standard prices are only used in a few cases where it is desirable to over-ride the average pricing method normally used for issues.

Stock Listings (Figs. 18 and 19)

A complete listing of the entire stock is produced each week and several copies are circulated to catering and production managers and storekeepers.

It is an invaluable reference and shows:

Stock number
Description
Packaging
Current average price (A) or standard price (S)
Date of last issue
Quantity in stock
Value of stock
Last supplier
Date of last receipt
Cost of last receipt
Cost of current production (production unit items only)

Classification number
An asterisk indicating items currently specified in recipes
Appropriate warning message.

Food Receipts Listing (Fig. 20)

The checking of invoices for payment with goods received is an important chore in any organisation. This listing helps the checking process and the checking ensures accuracy in the computer records.

Production Unit Listing (Fig. 21)

This listing itemises and gives the total value of each day's production. It can be checked against the original production records to confirm the accuracy of the computer records.

Food Issues Listing (Fig. 22)

This listing serves as an itemised invoice to the Catering Manager in each cost centre. It records the quantity and value of all food supplied.

Food Issues Summary (Fig. 23)

Perhaps the greatest advantage of a computer system is the ease with which summaries can be produced. In addition to the summary of a single week's food issues illustrated in Fig. 23, a number of other summaries are available to the same layout but covering year to date or specific trading periods (e.g., a term or a vacation period).

Recipe Costing (Figs. 24–28)

An extremely valuable by-product of the computerised costing system is the ability to produce accurate costings for recipes and menus. Figures 24, 25 and 26 illustrate recipes used in the Production Unit. Figure 27 illustrates a recipe which is used in individual kitchens to prepare a previously produced dish (i.e., loin of pork) for serving, while Fig. 28 is a recipe for an item produced directly in individual kitchens.

In all the recipes quantities may be given in either metric or imperial measure. Where imperial measures are used these are converted to metric. The average cost is taken from the costing records and is the current average issue price. For the Production Unit recipes a cost per pack is given. This is the cost which will be used for the pricing of current production of the item.

Fully detailed recipe costings are produced on request and also whenever there is a significant change in the ingredients specified as well as for all new recipes.

Recipe Cost Listings (Figs. 29 and 30)

The detailed recipe prints previously described contain cost information which quickly goes out of date. It would be wasteful to reprint these lists every time there was a price change. Likewise it would be dangerous to rely on out of date price information.

This problem is overcome by the production of an up to date list of recipe prices each week. Figure 29 illustrates kitchen recipes and Fig. 30 contains Production Unit recipes. Where the kitchen recipe uses a Production Unit pack as the major ingredient (sometimes the only ingredient) the pack size is shown in the column headed MULT.

Because these lists are produced weekly on the basis of current average food prices, up to date recipe costs are always available.

Menu Costing (Figs. 31 and 32)

It is but a short step, yet a complex one, from individual recipe costs to the costing of a complete multi-choice menu and from that to a complete cycle of menus covering a period of five weeks.

To achieve this, the Catering Manager must specify each of the dishes for a given menu on an input form R03. The computer will then calculate the costs and produce the listing shown in Fig. 31. This listing assumes a total of 1200 diners but this estimated total can be varied as can the split between the three refectories. Slight variations in cost will be noticed between the three refectories. This arises in part from variations in expected take-up of the dishes and in part from the inclusion of the cost of complete packs of food (i.e., three packs each serving eight people are needed to serve either 21 or 24 diners). Because it is a multi-choice menu, the computer lists the average cost per diner and also the maximum cost if a diner chooses the most expensive item in every course.

Again, in order to provide management with accurate menu costs without wasteful frequent reprinting of the full menu listings, we produce each week a summary of the five-weekly menu cycle as in Fig. 32. These costs are based on current average food prices thus ensuring that management always receives up to date costings.

01/11/79   UNIVERSITY OF KEELE - CATERING PRE-SORT VETTING   #CVET   DDE BATCH R01   2510   TIME 11/44   PAGE 1

```
  1   R02 SP0021417107916364      1RLIN                                                    REJECTED -  NEGATIVE TRANSFER INVALID
  2   R02 LIN31321210709383       2
FORM  R02      SHEET NO. 3132               QUANTITY DIFFERENCE    2     COST DIFFERENCE          NUMBER OF LINES      1      EXAMINE
  3   R01KHEDG15995900218107927206     5   3075 HAW
  4   R01KHEDG15995900218107999094    10   5880R                                              SHEET TOTAL
FORM  R01      SHEET NO. 9002               QUANTITY DIFFERENCE    5     COST DIFFERENCE  28.05   NUMBER OF LINES      2      EXAMINE
  5   R01KCEAR15983900918107950063    36   5136
  6   R01KCEAR15983900918107949096    36   3300
  7   R01KCEAR15983900918107949667    36   8424
  8   R01KCEAR15983900918107967254    36   2502
  9   R01KCEAR15983900918107968R64    48   2390
 10   R01KCEAR15983900918107936015   300  17400
 11   R01KCEAR15983900918107999994    90   2520                                              SHEET TOTAL
 12   R01KCEAR15983900918107999994   582  41672R
 13   R01KLATH15982884016107943520   127   8128
 14   R01KLATH15982884016107941633    18   1890
 15   R01KLATH15982884016107943697    23   1363
 16   R01KLATH15982884016107947034     2   1920                                              SHEET TOTAL
 17   R01KLATH15982884016107999994   170  13301R
 18   R01KWRIG15975884316107965779    36   2481                                              REJECTED -  INVALID DATE
 19   R01KWRIG15975884336107967328    48   3098
 20   R01KWRIG15975884316107956234    36   4712                                              WARNING -   DIRECT ISSUE INVALID
 21   R01KWRIG15975884316107942414     4   1861 HOR
 22   R01KWRIG15975884316107950056     3   2423
 23   R01KWRIG15975884316107948695     8   1852
 24   R01KWRIG15975884316107959122     4    801                                              REJECTED -  STOCK NUMBER INVALID
 25   R01KWRIG15975884316107967301     6    281
 26   R01KWRIG15975884316107955671     6   1109
 27   R01KWRIG15975884316107947229     2    115
 28   R01KWRIG15975884316107966476     2    475
 29   R01KWRIG15975884316107942894     2    493
 30   R01KWRIG15975884316107942894    12   1808
 31   R01KWRIG15975884316107942943     3   1808
 32   R01KWRIG15975884316107943432    72   1185
 33   R01KWRIG15975884316107959531    80   2017
 34   R01KWRIG15975884316107959764        3969
 35   R01KWRIG15975884316107999994   374  31388R
FORM  R01      SHEET NO. 8843               QUANTITY DIFFERENCE   52     COST DIFFERENCE  33.79   NUMBER OF LINES     16      EXAMINE
 36   R01KSTON15973R98215107921054    28   1680 HOR                                          SHEET TOTAL
 37   R01KSTON15973R98215107999994    28   1680R
```

Fig. 15.   Pre-sort vetting.

19/08/79   80/02 #CUPA

ITEM	PACKAGE	PRICE	TRANSACTION	QUANTITY	VALUE	SUPPLIER	SHEET	ISSUED TO	PAGE 40
66155 CARROTS	A10	0.709 A	CLOSING STOCK	126	89.30				0.928 C
66620 PEAS	A10	0.808 A	OPENING STOCK	62	50.08	CEAR			
			10/08 ISSUE	6	4.85		0117	SPO	
			14/08 ISSUE	6	4.85		0201	SPO	
			16/08 ISSUE	12	9.70		3102	LINDSAY	
		0.928 C	09/08 RECEIPT	36	33.42	15562 CEAR	8246		
		0.935 C	14/08 RECEIPT	24	22.44	15577 WRIG	8308		
		0.883 A	CLOSING STOCK	98	86.54				
67254 ROUND TOMATOES	3 KG	0.724 A	OPENING STOCK	28	20.27	WRIG			0.726 C
			14/08 ISSUE	12	8.69		3103	LINDSAY	
			15/08 ISSUE	18	13.03		2253	HORWOOD	
			16/08 ISSUE	2	1.45		3725	HAWTHORNS	
		0.724 C	14/08 RECEIPT	48	34.76	15577 WRIG	8308		
		0.724 A	CLOSING STOCK	44	31.86				
67328 MIXED VEGETABLES	A10	0.624 A	OPENING STOCK	13	8.11	SNOW			0.602 C
			10/08 ISSUE	6	3.74		3101	LINDSAY	
			14/08 ISSUE	12	7.49		3102	LINDSAY	
			15/08 ISSUE	18	11.23		2253	HORWOOD	
		0.645 C	07/08 RECEIPT	48	30.98	15554 WRIG	8233		
		0.645 C	14/08 RECEIPT	36	23.23	15577 WRIG	8308		
			ADJUSTMENT		0.51-				UP 7.1 %
		0.645 A	CLOSING STOCK	61	39.35				
68603 SALTED PEANUTS	24X4OZ	4.910 C	OPENING STOCK	5	24.55	CEAR			4.910 C
		5.330 C	09/08 RECEIPT	6	31.98	15600 CEAR	8319		
		5.139 A	CLOSING STOCK	11	56.53				UP 8.5 %

Fig. 16.   Transaction listing.

19/08/79 80/02 NCUPA   PAGE 50

	PACKAGE	PRICE	TRANSACTION	QUANTITY	VALUE	SUPPLIER	SHEET	ISSUED TO	PAGE
92607 CHICKEN CHASSEUR (8)	x 8	1.621 A	OPENING STOCK	89	164.30				1.621 C
			17/08 ISSUE	6	9.73		3100	LINDSAY	
		1.621 A	CLOSING STOCK	83	154.57				
92692 CURRIED CHICKEN	x 2	0.420 A	OPENING STOCK	1	0.42				0.432 C
			10/08 ISSUE	2	0.84		6423	MIS	
			10/08 ISSUE	1	0.42		6423	MIS	
		0.420 A	BALANCE C/FD	2-	0.84-				
EXAMINE BALANCE									
92815 CHICKEN & HAM PIE	x 6	0.679 A	OPENING STOCK	180	122.2R				0.679 C
			10/08 ISSUE	6	4.07		0117	SPO	
			17/08 ISSUE	5	3.40		0151	SPO	
			17/08 ISSUE	4	2.72		0151	SPO	
		0.679 A	CLOSING STOCK	165	112.09				
92856 CHICK MUSHROOM CASSEROLE	x 8	1.571 A	OPENING STOCK	78	122.56				1.571 C
			17/08 ISSUE	3	4.71		0151	SPO	
			17/08 ISSUE	18	28.28		2250	HORWOOD	
			17/08 ISSUE	25	39.28		3109	LINDSAY	
			17/08 ISSUE	33	51.84		3718	HAWTHORNS	
		1.571 C	15/08 PRODUCE	49	76.98	PRO	9999		
		1.571 A	CLOSING STOCK	48	75.41				
92942 ROAST CHICKEN	x 8	1.440 A	OPENING STOCK	77	110.88				1.440 C
			17/08 ISSUE	3	4.32		0151	SPO	
			17/08 ISSUE	10	14.40		2252	HORWOOD	
			17/08 ISSUE	11	15.84		2252	HORWOOD	
			17/08 ISSUE	32	46.08		3718	HAWTHORNS	
		1.440 C	14/08 PRODUCE	39	56.16	PRO	9999		
		1.440 A	CLOSING STOCK	60	86.40				

FIG. 17. Transaction listing.

PAGE 24

07/10/79 80/09 #CUPB	PACKAGE	PRICE	LAST MOVE	QUANTITY	VALUE	SUPPLIER	DATE	COST	PRODUCE	PAGE
49522 FRUIT COCKTAIL	3KG/A10	1.662 A	03/10/79	84	139.58	WRIG	28/08/79	1.375		17
49635 FRUIT SALAD	7# OZ	0.202 A	31/05/79	1	0.20	CEAR	12/01/79	0.203		17
49667 FRUIT SALAD	A10	2.017 A	12/09/79		6.19-	WRIG	21/08/79	2.017		17 *
49674 FRUIT COCKTAIL	7# OZ	0.175 A	28/09/79	6-	1.05-	CEAR	19/12/77	0.176		17
49716 GOOSEBERRIES	A10	1.551 A	24/09/79	96	148.86	WRIG	25/09/79	1.549		17 **
49762 GRAPEFRUIT SEGMENTS	A10	1.257 A	03/10/79	18	22.63	WRIG	04/09/79	1.265		17 *
49843 MANDARIN ORANGES	1LB14OZ	0.388 A	04/07/78			COLL	23/09/76	0.388		17
49875 MANDARIN ORANGES	3KG/A10	1.859 A	04/10/79	29	53.00	WRIG	11/09/79	1.815		17
50063 PEACH HALVES	A10	1.099 A	03/10/79	53	58.27	WRIG	25/09/79	1.054		17
50095 PEACH CUBES	A10	0.785 A	30/06/77			WRIG	05/04/77	0.785		17
50137 PEACH SLICES	3KG/A10	1.392 A	18/09/79	30	44.85	CEAR	07/09/79	1.648		17
50144 PEACH SLICES	7# OZ	0.296 A	02/02/79	14	4.15	CEAR	12/01/79	0.403		17 *
50225 PEARS	8 OZ	0.402 A	14/09/79	2	0.80	CEAR	12/01/79	0.403		17
50264 PEAR PIECES	A10	1.203 A	04/06/79	42	50.53	CEAR	11/09/77	1.145		17
50271 PEARS	3KG/A10	1.269 A	27/09/79	11	13.06	WRIG	16/09/77	1.389		17
50289 PRUNES	A10	1.372 A	03/10/79	5	6.86	WRIG	03/07/79	1.372		17 *
50377 PINEAPPLE RINGS	A10	1.259 A	14/09/79	37	46.60	SIDD	26/01/78	1.467		17
50384 PINEAPPLE PIECES	A10	1.343 A	16/12/78			WRIG	29/05/79	1.343		17
50440 PACK.PLUMS	A 10	1.537 A	04/06/79	24	36.89	CEAR	09/11/78	1.467		17 * CHECK USE
50458 GOLDEN PLUMS	A10	0.862 A	04/06/79			UNIT	17/02/77	0.862		17
50465 PLUMS – RED HALVES	A10	1.057 A	04/06/79					1.247		17
50472 PRUNES (LOCKWOODS)	A10									17
50539 RHUBARB	A10	0.773 A	28/09/79	12	9.28	CEAR	01/11/78	0.837		17
51003										38
51028 DRIED APRICOTS	LB	0.140 A	10/05/77							18
51081 CURRANTS	8 OZ	1.985 A	16/05/79	4	7.94	UNIT	28/06/78	2.940		18 * CHECK USE
51130 DATES	8 OZ	0.125 S	06/09/77							18
51162 FIGS	7 LB	0.407 A								18
51275 RAISINS	7 LB	2.849 A	28/09/79	8	22.79	WRIG	26/06/79	3.015		18
51331 SULTANAS	7 LB	2.971 A	27/08/79	12	35.65	WRIG	18/09/79	2.981		18 **
51500										38
51564 LEMON JUICE	1.2 LIT	1.917 A	04/10/79	60	115.00	RASJ	19/09/79	1.917		20
51571 GRAPEFRUIT JUICE	1.2 LIT	1.837 A	12/08/76	192	352.64	RASJ	19/09/79	1.917		20 *
51613 GRAPEFRUIT JUICE	26# OZ	0.260 A	29/04/77			DUNN	07/08/76	0.260		20
51620 ORANGE JUICE	34# OZ	0.300 A	04/10/79	288	556.68	DUNN	29/06/77	0.300		20
51677 ORANGE JUICE	1.2 LIT	1.933 A	11/11/76			RASJ	19/09/79	1.917		20
51719 ORANGE JUICE	A2	0.157 A	01/03/79			CEAR	12/08/76	0.155		20
51726 TOMATO JUICE	A5	0.296 A	09/04/79	24	13.87	WRIG	16/01/79	0.295		20
51740 TOMATO JUICE	A 10	0.578 A	27/07/79	6	12.22	WRIG	01/05/79	0.578		20 *
51797 TOMATO PUREE	5 KG	2.037 A				SNOW	27/08/79	2.037		27 * CHECK USE

FIG. 18. Stock listing.

Fig. 19. Stock listing.

30/09/79	80/08 #CUPB	PACKAGE	PRICE		LAST MOVE	QUANTITY	VALUE	SUPPLIER DATE	COST	PRODUCE		PAGE 49
97330	RATATOUILLE	x 6	0.664	A	24/08/79	4-	2.67-	22/11/78	0.664	0.672	03	*
97348	RATATOUILLE	x 2	0.221	A	10/08/79	100	22.10	05/01/79	0.221	0.224	03	
97387	RISSOTTO	x 6	0.325	A	31/08/79	34	11.03	04/05/79	0.331	0.331	03	
97394	RISSOTTO	x 3	0.162	A	08/06/79	56	9.07	30/11/78	0.161	0.166	03	
97450	NUTMEAT ROAST (IN BAG)	x 1	0.054	A	26/01/78					0.078	03	*
97468	RISSOTTO MILANESE	x 6	0.912	A				26/04/77	0.912	0.892	03	* CHECK USE
97475	SAVOURY ROAST (NUTMEAT)	x 8	0.530	A						0.627	03	* CHECK USE
97482	SEPTEMBER STEW	x 8	1.239	A	13/11/76			31/05/79	1.250	1.242	03	
97517	SEPTEMBER STEW	x 2	0.315	A	24/06/79	33	10.46	13/06/79	0.312	0.310	03	
97524	SPINNACH PIE	x 6	0.424	A	21/09/79	24	10.18	30/09/77	0.434	0.468	03	
97553	SWEETCORN CROQUETTES	x 8	0.568	A	19/06/79	21	11.92	24/01/79	0.569	0.570	03	*
97605	VEGETABLE CASSEROLE	x 0	1.058	A	14/09/79			20/08/79	1.058	1.058	03	
97612	VEGETABLE CASSEROLE	x 2	0.353	S	13/07/79			27/04/79	0.352	0.353	03	
97637	VEGETABLE PIE	x 6	0.258	S		12	4.25			0.236	04	
97644	VEGETABLE CURRY	x 2	0.325	A	14/09/79	60	1.30	22/06/79	0.325	0.323	04	
97651	VEGETABLE CURRY	x 8	0.909	A	31/08/79	9	8.18	22/06/79	0.976	0.968	04	
97690	VEGETABLE CUTLETS	x 10	0.316	A	14/09/79	14	4.42	12/09/79	0.316	0.316	04	
98006	* APPLE BLACKBERRY TART	x 1	0.303	A	10/08/79	24	7.32	02/08/79	0.288	0.005	04	*
98013	APPLE BLACKCURRANT PIE	x 8	0.319	A	28/09/79	71	22.65	20/09/79	0.320	0.307	04	
98052	APPLE CHARLOTTE	x 8	0.337	A	18/07/79	224	75.49	15/06/79	0.342	0.318	04	
98101	APPLE CHARLOTTE	x 8	0.327	A	31/08/79	115	37.50	05/09/79	0.327	0.304	04	
98140	APPLE CRUMBLE	x 10	0.340	A	21/09/79	84	28.54	13/09/79	0.340	0.321	04	
98197	APPLE DUMPLING	x 8	0.273	A	28/09/79	54	14.72	14/09/79	0.280	0.262	04	
98239	APPLE TART	x 8	0.300	A	28/09/79	106	31.78	21/09/79	0.304	0.286	04	
98278	DUTCH APPLE TART	x 8	0.445	A	28/09/79	73	32.50	27/09/79	0.430	0.428	04	
98366	APRICOT CRUMBLE	x 10	0.500	A	13/07/79	59	29.48	27/09/79	0.488	0.490	04	
98408	APRICOT DUMPLING	x 8	0.359	A	28/09/79	235	84.27	23/08/79	0.359	0.362	04	
98454	APRICOT TART	x 8	0.221	A	24/02/78			11/02/77	0.274	0.383	04	**
98493	APRICOT UPSIDE DOWN PUDD	x 8	0.220	A	21/09/79	203	44.66	26/09/79	0.220	0.221	04	*
98535	BAKEWELL TART	x 8	0.378	A	22/06/79	226	85.43	15/05/79	0.372	0.378	04	**
98581	BREAD & BUTTER PUDDING	x 10	1.019	S	30/06/77					1.302	04	
98662	CHARLOTTE RENAISSANCE	x 10	0.541	S	21/09/79	44	52.49	21/06/79	1.301	1.975	04	
98711	CHARLOTTE RUSSE	x 10	1.193	A	28/01/77					1.300	04	
98717	CHERRY TART	x 10	0.390	A	28/09/79	105	33.52	25/09/79	0.318	0.782	04	
98824	CHERRY DUMPLING	x 8	0.245	A	03/08/79	253	62.06	05/09/79	0.252	0.318	04	*
98856	CHOCOLATE & FRUIT SPONGE	x 8	0.226	A	16/02/79					0.252	04	
98870	COCONUT SPONGE	x 4	0.124	A	28/06/77	7	0.87			0.674	04	
98905	CRANBERRY TART	x 8	0.659	A	28/09/79	55	36.25	20/09/79	0.659	0.216	04	NOT MOVED
98944	CREPE SUZETTE									0.663	04	
98951	COFFEE CRUNCH											

19/08/79   80/02 #CREP   UNIVERSITY OF KEELE –   FOOD RECEIPTS UNDER KEELE DIRECT ORDERS –   CATERING   PAGE 5

ORDER NUMBER	REFERENCE OR DATE	SHEET NUMBER	SUPPLIER CODE	DESCRIPTION	STOCK NUMBER	QUANTITY	HASH TOTAL	COST	SHEET COST	ORDER COST
15564	14/08/79	8303	SNOW	SALAD CREAM (IND)	5711R	5		19.15		
15564	14/08/79	8303	SNOW	HONEY (PORTIONS)	58390	3		18.87		
15564	14/08/79	8303	SNOW	MARMALADE	58489	10		38.90		
15564	14/08/79	8303	SNOW	ASSORTED JAMS	59115	5		20.55		
15564	14/08/79	8303	SNOW	COOKING OIL	59001	6		53.70		
15564	14/08/79	8303	SNOW	TOMATO SAUCE (IND)	60509	5		19.15		
15564				TOTAL ORDER VALUE			34		170.32	170.32
15566	09/08/79	8236	HEDG	CHIPS (FROZEN)	27218	4		22.44		
15566	09/08/79	8236	HEDG	CHIPS (FROZEN)	27218	5		28.05		
15566	09/08/79	8236	HEDG	POTATO CROQUETTES	27271	3		11.10		
15566	09/08/79	8236	HEDG	POTATO CROQUETTES	27271	2		7.40		
15566	09/08/79	8236	HEDG	CHICKEN PURCHASES	38100	900		276.00		
15566				TOTAL ORDER VALUE			914		344.99	344.99
15571	10/08/79	8194	FMCM	LAMB PURCHASES	34018	343		246.96		
15571	10/08/79	8194	FMCM	LAMB PURCHASES	34018	53	396	26.50	273.46	
15571	14/08/79	8192	FMCM	BEEF PURCHASES	30006	372		374.00		
15571	14/08/79	8192	FMCM	PORK PURCHASES	36015	145	517	36.25	410.25	
15571	15/08/79	8193	FMCM	PORK PURCHASES	36015	137	137	34.25	34.25	
15571				TOTAL ORDER VALUE						717.96
15572	10/08/79	8239	LYON	ICE CREAM SOFT SCOOP	24143	20		27.20		
15572	10/08/79	8239	LYON	ICE CREAM SOFT SCOOP	24143	20		27.20		
15572	10/08/79	8239	LYON	VAT KEELE	29945	1	41	12.24	66.64	
15572				TOTAL ORDER VALUE						66.64
15573	10/08/79	8185	SUNB	HAMBURGERS	28155	4		5.40		
15573	10/08/79	8185	SUNB	SAUSAGE PURCHASES	38904	160	164	45.60	51.00	
15573	10/08/79	8186	SUNB	REGAL STEAKBURGERS	28123	2	2	9.90	9.90	
15573	10/08/79	8187	SUNB	SAUSAGE PURCHASES	38904	160	160	45.60	45.60	
15573				TOTAL ORDER VALUE						106.50

Fig. 20.   Food receipts listing.

19/08/79 80/02 #CREP   UNIVERSITY OF KEELE -   CATERING DEPARTMENT - PRODUCTION UNIT - PRODUCTION   PAGE 1

DATE ISSUED	REQUISITION NUMBER	STOCK NUMBER	DESCRIPTION	ISSUE PRICE	QUANTITY	COST	MASH TOTAL	SHEET COST	LAST YEAR
10/08/79	9999	93067	COTTAGE PIE	0.806	26	20.96			
10/08/79	9999	93706	LIVER & ONIONS BRAISED	1.251	16	20.02			
10/08/79	9999	92163	MINCED BEEF & ONIONS	0.735	10	7.35			
10/08/79	9999	92205	ROAST BEEF	2.151	82	176.38			
10/08/79	9999	93649	ROAST LEG OF LAMB	2.642	47	124.17			
10/08/79	9999	92290	SAUTE BEEF & CARROTS	1.590	39	62.01			
10/08/79	9999	94996	STUFFING BALLS	0.284	24	6.82			
10/08/79	9999	95340	YORKSHIRE PUDDING (9)	0.132	56	7.39	300	425.10	
14/08/79	9999	90150	FISH DUGLERE	1.509	19	30.38			
14/08/79	9999	99257	GOOSEBERRY TART	0.486	82	39.85			
14/08/79	9999	92205	ROAST BEEF	2.151	47	101.10			
14/08/79	9999	92942	ROAST CHICKEN	1.440	39	56.16			
14/08/79	9999	94851	STEAK & KIDNEY PUDDING	0.834	39	32.53			
14/08/79	9999	94996	STUFFING BALLS	0.284	17	4.83	243	264.85	
15/08/79	9999	98052	APPLE BLACKCURRANT PIE	0.322	112	36.06			
15/08/79	9999	92063	CARBONADE OF BEEF	1.396	20	27.92			
15/08/79	9999	92854	CHICK MUSHROOM CASSEROLE	1.571	49	76.98			
15/08/79	9999	90818	FISH WALESKA	2.032	17	34.54			
15/08/79	9999	90384	MINCEMEAT TART	0.323	89	28.75			
15/08/79	9999	93649	ROAST LEG OF LAMB	2.642	86	227.21	373	431.46	
16/08/79	9999	98239	APPLE TART	0.284	120	34.08			
16/08/79	9999	92074	COQ AU VIN	1.096	47	93.81			
16/08/79	9999	90362	FISH MORNAY (8)	1.693	18	30.67			
16/08/79	9999	94227	LOIN OF PORK	2.453	35	85.86			
16/08/79	9999	92290	SAUTE BEEF & CARROTS	1.500	31	49.20			
16/08/79	9999	95340	YORKSHIRE PUDDING (9)	0.132	58	7.66	309	301.17	

TOTAL COST OF ITEMS IN THIS SCHEDULE   1,422.58

FIG. 21. Production unit listing.

DATE ISSUED	REQUISITION NUMBER	STOCK NUMBER	DESCRIPTION	ISSUE PRICE	QUANTITY	COST	HASH TOTAL	SHEET COST	LAST YEAR
23/02/79	3610	94530	SPAGHETTI BOLOGNAISE	0.420	4	1.68			
23/02/79	3610	94717	SPANISH BAKE	0.984	5	4.92			
23/02/79	3610	94932	STEAK & ONION PIE	0.450	8	3.67			
23/02/79	3610	94996	STUFFING BALLS	0.275	4	1.10			
23/02/79	3610	95037	TOAD IN THE HOLE	0.532	6	3.19			
23/02/79	3610	97690	VEGETABLE CUTLETS	0.531	1	0.31	114	98.26	
23/02/79	3611	98140	APPLE CRUMBLE	0.329	12	3.95			
23/02/79	3611	98454	APRICOT TART	0.329	18	5.92			
23/02/79	3611	92124	BEEF HOTPOT	1.243	8	9.94			
23/02/79	3611	98856	CHOCOLATE & FRUIT SPONGE	0.316	12	3.79			
23/02/79	3611	92068	CURRIED BEEF	1.297	6	7.78			
23/02/79	3611	96658	EGGS FLORENTINE	0.340	2	0.68			
23/02/79	3611	90127	FISH BRETONNE	1.500	1	1.50			
23/02/79	3611	90215	FISH FLORENTINE	1.880	1	1.88			
23/02/79	3611	90342	FISH MORNAY (8)	1.670	1	1.67			
23/02/79	3611	99516	PINEAPPLE UPSIDE DOWN	0.418	18	7.52			
23/02/79	3611	97200	PIZZA	0.215	2	0.43			
23/02/79	3611	92205	ROAST BEEF	2.150	20	43.00			
23/02/79	3611	92942	ROAST CHICKEN	1.440	10	14.40			
23/02/79	3611	97482	SEPTEMBER STEW	1.210	1	1.21			
23/02/79	3611	96800	STEAK & MUSHROOM PIE	0.495	6	2.97			
23/02/79	3611	97235	STUFFED PEPPERS	0.235	2	0.47			
23/02/79	3611	94996	STUFFING BALLS	0.273	3	0.82			
23/02/79	3611	95340	YORKSHIRE PUDDING (9)	0.163	18	2.93	141	110.95	
23/02/79	3612	98052	APPLE BLACKCURRANT PIE	0.317	16	5.07			
23/02/79	3612	96136	CABBAGE CASSEROLE	0.175	2	0.35			
23/02/79	3612	92646	CHICKEN CROQUETTES	0.553	3	1.66			
23/02/79	3612	96545	CHINESE VEGETABLES	0.180	2	0.36			
23/02/79	3612	96577	DHALL	0.130	1	0.13			
23/02/79	3612	90085	FISH BONNE FEMME (8)	1.560	1	1.56			
23/02/79	3612	99112	GOOSEBERRY CHARLOTTE	0.480	12	5.76			
23/02/79	3612	99384	MINCEMEAT TART	0.317	12	3.80			
23/02/79	3612	92942	ROAST CHICKEN	1.440	2	2.88			
23/02/79	3612	94530	SPAGHETTI BOLOGNAISE	0.420	2	0.84			
23/02/79	3612	94851	STEAK & KIDNEY PUDDING	0.803	4	3.21			
23/02/79	3612	94996	STUFFING BALLS	0.270	1	0.27	58	25.89	
25/02/79	0000	09904	DRINK STOCK PRICE ADJUST		0	0.02			
25/02/79	0000	09904	FOOD STOCK PRICE ADJUST		0	2.48			

25/02/79   79/29 MCREP   UNIVERSITY OF KEELE – CATERING DEPARTMENT – FOOD ISSUES   HAWTHORNS   PAGE 6

Fig. 22. Food issues listing.

07/10/79 80/09 NCUPA — UNIVERSITY OF KEELE — FOOD ISSUES SUMMARY — WEEK ENDED 23/09/79 — SHEET 2

CLASS OF FOOD	MORWOOD	%	LINDSAY	%	HAWTHORNS	%	SUB-TOTAL	%	PRODUCTION	%	OTHERS	%	TOTAL ISSUES	%
01 BLAST FREEZE - FISH DISH	107.44	3.8	97.48	8.3	37.36	3.7	242.28	4.9					242.28	3.5
02 BLAST FREEZE - MEAT DISH	465.14	16.7	232.34	19.9	258.55	25.8	956.03	19.3			172.72	28.5	1128.75	16.5
03 BLAST FREEZE - VEGETARIA			14.77	1.2			14.77	0.3					15.06	0.2
04 BLAST FREEZE - SWEETS	87.50	3.1	55.81	4.8	58.02	5.8	201.33	4.0			5.75	0.9	207.08	3.0
05 BEVERAGES	78.64	2.8	26.21	2.2			104.85	2.1			53.32	8.8	158.17	2.3
06 BISCUIT CRISPS CHOCOLATE	38.46	1.3	17.88	1.5	11.92	1.1	68.26	1.4			13.92	2.3	82.17	1.2
07 BREAD CAKES PIES	13.40	0.4	20.10	1.7	6.70	0.6	40.20	0.8			2.31	0.3	42.51	0.6
08 BREAKFAST CEREALS					3.24	0.3	3.24						3.24	
09 CHEESE	84.95	3.0	51.58	4.4	13.13	1.3	149.66	3.0			8.99	1.4	158.65	2.3
10 CONDIMENTS	33.49	1.2	8.35	0.7	8.78	0.8	50.62	1.0	19.04-		4.82	0.8	36.40	0.5
11 DAIRY PRODUCE - MILK ETC	120.84	4.3	27.60	2.3	127.89	12.8	276.33	5.6			29.73	4.9	306.06	4.4
12 EGGS	61.20	2.2	17.65	1.5	9.60	0.9	88.45	1.7			21.37	3.5	109.82	1.6
13 FISH - CANNED														
14 FISH - FRESH	74.64	2.6	78.35	6.7	83.02	8.3	236.01	4.7	157.50	12.1	5.75	0.9	399.26	5.8
15 FISH - FROZEN														
16 FARINACEOUS PRODUCTS	15.84	0.5			10.65	1.0	26.49	0.5					26.49	0.3
17 FRUIT - CANNED	23.59	0.8	8.35	0.7	20.70	2.0	52.64	1.0					52.64	0.7
18 FRUIT - DRIED														
19 FRUIT - FRESH	57.48	2.0	5.43	0.4	6.42	0.6	69.33	1.4			2.43	0.4	71.76	1.0
20 FRUIT JUICES & MINERALS	52.00	1.8					52.00	1.0					52.00	0.7
21 ICE CREAM	94.61	3.4	39.69	3.4	16.70	1.4	149.00	3.0			12.12	2.0	161.12	2.3
22 MEAT	667.95	24.0	182.04	15.6	178.81	17.9	1028.80	20.8	1081.33	83.6	3.83	0.6	2113.96	30.9
23 MEAT - BACON	39.38	1.4					39.38	0.8			19.69	3.2	59.07	0.8
24 MEAT - CANNED	31.14	1.1	21.55	1.8	13.70	1.3	66.39	1.3			8.81	1.4	75.20	1.1
25 MIXES - CUSTARD & GRAVY	6.33	0.2			17.34	1.7	23.67	0.4					23.67	0.3
26 MIXES - OTHERS					2.02	0.2	2.02						9.20	0.1
27 PICKLES & SAUCES	4.03	0.1					9.20	0.1			3.15	0.5	150.20	2.2
28 POTATO	88.00	3.1	28.60	2.4	33.60	3.3	150.20	3.0					7.97	0.1
29 POTATO - MIXES			7.90	0.6			7.97	0.1					19.12	0.2
30 PRESERVES - JAMS ETC	13.32	0.4	54.72	4.7	42.58	4.2	19.12	0.3					257.34	3.7
31 PROVISIONS - BUTTER ETC	149.28	5.3	9.14	0.7	20.56	2.0	246.58	4.9			10.76	1.7	52.34	
32 SOUPS	22.64	0.8	25.51	2.1			49.75	1.0	61.30	4.7	22.19	3.6	133.33	
33 SUGAR	24.24	0.8	15.93	1.3			27.54	0.5			10.81	1.7	38.35	0.5
34 VEGETABLES - CANNED	11.61	0.4	10.75-				23.91	0.4	15.39	1.1	120.17	19.8	159.47	2.3
35 VEGETABLES - FRESH	37.47	1.3	108.47	9.3	2.81-		375.60	7.6			71.36	11.7	446.96	6.5
36 VEGETABLES - FROZEN	258.82	9.3			8.21	0.8							8.19-	
37 VEGETARIAN ITEMS									8.19-					
38 PRICE ADJUSTMENTS	11.38	0.4	11.84	1.0	14.35	1.4	37.57	0.7	7.21	0.5			44.78	0.6
39 ** TOTAL FOOD ISSUES **	2775.11		1162.41		999.04		4936.56		1295.59		605.19		6837.36	
40 BEER SPIRIT WINE TOBACCO	14.00-		70.16		20.29		55.55		12.90		30.65		119.10	

FIG. 23. Food issues summary.

07/10/79   80/09 MCREP   UNIVERSITY OF KEELE -   RECIPE COSTING   W/E 07/10/79   SHEET 45

STOCK NUMBER	DESCRIPTION		METRIC QUANTITY	AVERAGE COST	SERVES	COST PER SERVING	SERVINGS PER PACK	COST PER PACK
90085	FISH BONNE FEMME (8)				96			
01 21047	FISH (COD)	28 LB	12.701	19.026				
02 19245	MUSHROOMS	3 LB	1.361	2.298				
03 59764	DAWN MARGARINE	8 OZ	0.227	0.100				
04 47229	COOKING SALT		0.070	0.003				
05 47067	GROUND WHITE PEPPER		0.010	0.039				
06 52254	MIXED HERBS		0.010	0.031				
07 19421	ONIONS	2 LB	0.907	0.109				
08 14134	MILK	5 PINT	2.841	0.750				
09 89921	CHARBONNIER WHITE		1.000	1.112				
10 18354	LEMONS	4 OZ	0.113	0.080				
11 42414	FLOUR	5 OZ	0.142	0.020				
12 42661	BADEX STARCH	8 OZ	0.227	0.150				
13 15353	CREAM - DOUBLE	1 PINT	0.568	1.065				
14 59531	BUTTER	2 OZ	0.057	0.090				
	RECIPE TOTALS			24.063	96	0.260	8	2.080

Fig. 24. Recipe costing.

07/10/79   80/09 #CREP   UNIVERSITY OF KEELE   -   RECIPE COSTING   W/E 07/10/79   SHEET 57

STOCK NUMBER	DESCRIPTION		METRIC QUANTITY	AVERAGE COST	SERVES	COST PER SERVING	SERVINGS PER PACK	COST PER PACK
93833	MEAT & POTATO PIE				846			
01 31200	CUBED BEEF	50 LB	22.680	29.982				
02 19580	PEELED POTATOES	30 LB	13.608	2.340				
03 42616	FLOUR	1 LB	0.454	0.066				
04 42661	BADEX STARCH	1 LB	0.454	0.700				
05 42414	FLOUR	65 LB	29.484	4.304				
06 59764	DAWN MARGARINE	16 LB 4 OZ	7.371	3.272				
07 59683	LARD	16 LB 4 OZ	7.371	3.235				
08 47902	EGGS		2.000	0.050				
09 19421	ONIONS	10 LB	4.536	0.097				
10 47229	COOKING SALT		0.592	0.027				
11 47067	GROUND WHITE PEPPER		0.010	0.039				
12 42083	BAKING POWDER	1 LB 8 OZ	0.680	0.373				
13 14134	MILK	7 FO	0.199	0.052				
	RECIPE TOTALS			45.037	846	0.053	6	0.319

FIG. 25. Recipe costing.

07/10/79   80/09 #CREP   UNIVERSITY OF KEELE   -   RECIPE COSTING   W/E 07/10/79   SHEET 95

STOCK NUMBER	DESCRIPTION		METRIC QUANTITY	AVERAGE COST	SERVES	COST PER SERVING	SERVINGS PER PACK	COST PER PACK
98366	APRICOT CRUMBLE				560			
01 49272	APRICOTS	97 LB	43.999	18.919				
02 65024	SUGAR	12 LB	5.643	1.834				
03 42661	BADEX STARCH	2 LB 8 OZ	1.134	0.749				
04 42614	FLOUR	45 LB	20.412	2.080				
05 59764	DAWN MARGARINE	22 LB 8 OZ	10.206	4.531				
06 65024	SUGAR	15 LB	6.804	2.292				
07 42083	BAKING POWDER	1 LB 8 OZ	0.680	0.373				
08 47229	COOKING SALT		0.090	0.004				
	RECIPE TOTALS			31.682	560	0.056	8	0.452

FIG. 26.   Recipe costing.

07/10/79   80/09 #CREP   UNIVERSITY OF KEELE   -   RECIPE COSTING   W/E 07/10/79   SHEET 9

STOCK NUMBER	DESCRIPTION		METRIC QUANTITY	AVERAGE COST	SERVES	COST PER SERVING	SERVINGS PER PACK	COST PER PACK
03604	ROAST PORK & APPLE SAUCE				90			
01 94227	LOIN OF PORK	6 LB	9.000	22.077				
02 40096	PACK APPLES		2.722	0.090				
03 59764	DAWN MARGARINE	4 OZ	0.113	0.050				
04 65024	SUGAR	1 LB	0.454	0.152				
05 55424	GRAVY MIX	11 OZ	0.312	0.315				
06 19301	CRESS		0.090	0.225				
	RECIPE TOTALS			23.809	90	0.264		

FIG. 27.   Recipe costing.

07/10/79   80/09 MCREP   UNIVERSITY OF KEELE   -   RECIPE COSTING   W/E 07/10/79   SHEET 20

STOCK NUMBER	DESCRIPTION		METRIC QUANTITY	AVERAGE COST	SERVES	COST PER SERVING	SERVINGS PER PACK	COST PER PACK
06179	CAULIFLOWER MORNAY				100			
01 19118	CAULIFLOWER	24 LB	10.886	2.394				
02 14134	MILK	16 PINT 7 FO	9.291	0.452				
03 50764	DAWN MARGARINE	2 LB 1 OZ	0.936	0.415				
04 42614	FLOUR	2 LB 1 OZ	0.936	0.136				
05 19621	ONIONS	8 LB 4 OZ	3.742	0.823				
06 64091	GROUND CLOVES		0.167	1.254				
07 43520	CHEDDAR CHEESE	4 LB 2 OZ	1.871	2.550				
08 47229	COOKING SALT		0.050	0.002				
09 47067	GROUND WHITE PEPPER		0.010	0.039				
10 46761	MUSTARD		0.417	0.479				
	RECIPE TOTALS			10.544	100	0.105		

Fig. 28. Recipe costing.

19/08/79	80/02 #CUPB	RECIPE	SERVES	COST	MULT	PER PORTION	PER PACK
02625	PRAWN PATTIES		93	11.73	1	0.126	
02664	FISH CAKES & PARSLEY SAUCE		100	5.04	1	0.050	
02689	SALMON FISH CAKES PARSLEY SAUCE		100	7.10	1	0.071	
02696	SALMON FISH CAKES TOMATO SAUCE		45	3.43	1	0.076	
02720	SALMON CROQUETTES SHRIMP SAUCE		100	10.32	1	0.103	
02960	FISH (BREAKFAST)		100	10.68	1	0.107	
03001	ROAST BEEF & HORSERADISH SAUCE		80	19.48	10	0.244	
03019	ROAST BEEF & YORKSHIRE PUDDING		90	21.08	10	0.234	
03026	BEEF HOT POT		8	1.25	8	0.157	
03033	CARBONADE OF BEEF		8	1.39	8	0.175	
03040	CURRIED BEEF & RICE		8	1.46	8	0.183	
03058	SAUTED BEEF & CARROTS		8	1.59	8	0.199	
03065	SAUTE BEEF & MUSHROOMS		8	1.59	8	0.199	
03097	BRAISED STEAK & CARROTS		10	2.51	10	0.251	
03114	BRAISED STEAK & DUMPLINGS		10	2.60	10	0.260	
03139	CHIPPED RUMP STEAK & TOMATO		350	91.11	1	0.126	
03153	BREADED VEAL CUTLETS		20	0.07	1	0.077	
03192	CHICKEN A LA KING		8	1.76	8	0.221	
03202	ROAST CHICKEN BREAD SAUCE BACON		96	21.25	8	0.221	
03234	ROAST CHICKEN BREAD SAUC STUFFIN		96	20.52	8	0.214	
03259	CASSEROLE CHICKEN & MUSHROOMS		8	1.58	8	0.190	
03273	CHICKEN CHASSEUR		8	1.63	8	0.205	
03315	CHICKEN CROQUETTES		8	0.70	8	0.088	
03347	CURRIED CHICKEN & RICE		8	1.94	8	0.243	
03379	COQ AU VIN		8	2.00	8	0.250	
03403	ROAST TURKEY HAM BREAD & STUFFNG		100	26.64	10	0.266	
03509	ROAST LEG OF LAMB & MINT SAUCE		176	79.12	8	0.450	
03523	ROAST LAMB & REDCURRANT JELLY		8	3.67	8	0.450	
03530	LAMB CHOP GARNI		30	7.31	1	0.244	
03548	LAMB CUTLETS REFORME		100	25.61	1	0.256	
03555	ROAST SADDLE OF LAMB		120	57.06	1	0.476	
03587	LUNCHEON MEAT FRITTERS		20	0.78	1	0.039	
03594	MOUSSAKA		8	1.89	8	0.236	
03604	ROAST PORK & APPLE SAUCE		90	23.69	10	0.263	
03629	ROAST PORK APPLE SAUCE & STUFFNG		240	65.73	10	0.274	
03643	GRILLED PORK CHOP GARNI		20	3.87	1	0.194	
03668	GRILLED PORK CHOP & PIQUANT SAUC		160	31.36	1	0.196	
03675	ROAST POUSSIN		96	63.81	8	0.665	
03682	GAMMON & PINEAPPLE		15	5.55	1	0.370	
03724	GRILLED GAMMON & SWEETCORN		10	2.78	1	0.279	

FIG. 29. Recipe cost listing.

SHEET 6

19/08/79	80/02 #CUPB RECIPE	SERVES	COST	MULT	PER PORTION	PER PACK
98969	CHOCOLATE & ORANGE CRUNCH	400	29.29	8	0.073	0.586
98983	EVES PUDDING	312	15.84	8	0.051	0.406
98990	RHUBARB EVE(S PUDDING	312	13.85	8	0.044	0.355
99031	FRUIT SUET ROLL (DUMPLG)	400	10.55	10	0.026	0.264
99070	FRUIT SPONGE	312	10.94	8	0.035	0.281
99112	GOOSEBERRY CHARLOTTE	440	29.83	8	0.068	0.542
99144	GOOSEBERRY CRUMBLE	560	32.29	8	0.058	0.461
99183	GOOSEBERRY EVES PUDDING	312	17.88	8	0.057	0.458
99257	GOOSEBERRY TART	320	19.49	8	0.061	0.487
99296	JAM SPONGE	312	12.25	8	0.039	0.314
99313	LEMON SPONGE	312	8.99	8	0.029	0.231
99338	MARMALADE SPONGE	312	12.40	8	0.040	0.318
99377	MINCE PIES	590	10.85	12	0.018	0.221
99384	MINCEMEAT TART	320	12.93	8	0.040	0.323
99426	PASTRY CASES	320	3.56	8	0.011	0.089
99440	PEACH TART	320	20.47	8	0.064	0.512
99465	PEAR UPSIDE DOWN	312	16.74	8	0.054	0.429
99514	PINEAPPLE UPSIDE DOWN	312	16.56	8	0.053	0.425
99553	PLUM CRUMBLE	560	23.94	8	0.043	0.342
99592	PLUM SPONGE	312	12.16	8	0.039	0.312
99641	PLUM TART	320	13.27	8	0.041	0.332
99680	RHUBARB GINGER CRUMBLE	560	20.32	8	0.036	0.290
99722	RHUBARB GINGER TART	320	8.44	8	0.026	0.211
99779	RICE PUDDING	96	3.40	8	0.036	0.284
99867	PLAIN SPONGE	312	8.20	8	0.026	0.211
99909	STEAMED GINGER SPONGE	312	12.41	8	0.040	0.318
99016	SYRUP SPONGE	280	8.40	8	0.030	0.240
99948	SYRUP TART	560	15.87	8	0.028	0.227
99955	MELON & GRAPEFRUIT COCKT	432	37.51	12	0.087	1.042
99962	MELON[ORANGE COCKTAIL	374	19.81	12	0.053	0.636

FIG. 30. Recipe cost listing.

28/10/79 80/12 MCREP UNIVERSITY OF KEELE - MENU 325 - TUESDAY DINNER - SPECIMEN COST SHEET 2

DISH NUMBER	ITEM NUMBER	COURSES AND DISHES	MORKWOOD SERVE	MORKWOOD COST	LINDSAY SERVE	LINDSAY COST	HAWTHORNS SERVE	HAWTHORNS COST	TOTAL SERVE	TOTAL COST	COST PER SERVING
01903	11	FRUIT JUICES	247	5.820	196	4.570	146	3.486	589	13.876	0.023
01420	12	TOMATO SOUP	253	7.073	206	5.761	152	4.251	611	17.085	0.027
03379	21	COQ AU VIN	212	63.207	166	49.161	124	37.456	502	149.824	0.298
02590	22	FISH BONNE FEMME	51	14.574	40	10.410	33	10.410	124	35.394	0.285
06326	23	MACARONI AU GRATIN	40	2.576	24	1.472	21	1.472	85	5.520	0.064
03139	24	CHIPPED RUMP STEAK & TOM	141	20.054	121	17.209	88	12.515	350	49.778	0.142
05030	25	CHEESE & NUT PANCAKES	5	0.505	8	1.010	6	0.505	19	2.020	0.106
05584	26	PIZZA	51	3.224	40	2.480	30	1.984	121	7.688	0.063
06362	31	TOMATOES	288	14.129	234	11.479	173	8.487	695	34.095	0.049
06228	32	MACEDOINNE	212	5.481	166	4.291	127	3.283	505	13.055	0.025
06813	41	PARSLEY POTATOES	152	3.819	117	2.938	94	2.360	363	9.117	0.025
06852	42	SAUTE POTATOES	348	7.868	283	6.382	206	4.645	837	18.875	0.022
08672	51	STRAWBERRY MOUSSE & CREA	101	1.547	81	1.257	61	0.946	243	3.770	0.015
07285	52	BAKEWELL TART & CUSTARD	197	8.968	158	7.159	118	5.360	473	21.476	0.045
07510	53	FRESH FRUIT	202	11.371	162	9.119	121	6.811	485	27.301	0.056
09518	61	CHEESE & BISCUITS	455	28.077	364	22.462	273	16.844	1092	67.383	0.061
09010	62	COFFEE/TEA	45	3.145	36	2.521	27	1.882	108	7.548	0.060
		TOTAL DINERS & COSTS	500	201.418	400	150.681	300	122.706	1200	483.805	
		AVERAGE COST PER DINER		0.403		0.399		0.409		0.403	
		MAXIMUM COST PER DINER		0.524		0.523		0.541		0.524	

FIG. 31. Menu costing.

15/07/79   79/49 #CUPB   ESTIMATED CURRENT FOOD COSTS PER 100 DINERS PER MEAL   SHEET A

	WEEK 1	2	3	4	5	AVERAGE	7	8	AVERAGE
LUNCHEON - MONDAY	32.31	35.18	31.08	34.74	35.61		39.45	37.26	
TUESDAY	33.24	38.47	32.87	28.95	30.47		40.60	43.02	
WEDNESDAY	34.72	33.77	31.21	33.53	35.46		45.57	45.40	
THURSDAY	33.17	28.37	33.41	31.10	33.83		44.87	31.53	
FRIDAY	34.06	34.74	35.81	33.02	34.59		34.41	43.03	
SATURDAY							34.56	43.05	
SUNDAY							53.21	53.37	
TOTAL FOR WEEK	167.50	170.53	164.38	162.24	169.96	166.92	292.65	304.96	298.81
AVERAGE DAILY COST	33.50	34.11	32.88	32.45	33.99	33.39	41.81	43.57	42.69
DINNER - MONDAY	49.56	37.76	54.35	32.31	61.37		53.06	66.32	
TUESDAY	40.82	41.93	36.21	48.90	47.16		82.00	50.47	
WEDNESDAY	28.78	44.52	52.51	45.83	42.72		50.08	51.85	
THURSDAY	42.16	34.73	41.62	43.02	43.66		49.57	50.94	
FRIDAY	35.92	35.78	49.07	41.25	49.32		64.75	71.06	
SATURDAY							46.06	45.85	
SUNDAY							44.98	44.74	
TOTAL FOR WEEK	107.24	194.72	233.76	211.31	240.23	215.45	390.48	381.23	385.86
AVERAGE DAILY COST	39.45	38.94	46.75	42.26	48.05	43.09	55.78	54.66	55.12
AVERAGE BREAKFAST COST	26.22								
MORNING COFFEE & BISC.	5.80								
AFTERNOON TEA & BISC.	4.10								

Fig. 32.   Menu costing.

Index

Ardeer, ICI factory, 5

Biological checks, 23

Catering
  conventional system, 5, 45
  cost control, 38
  costing system, 157–95
    computer application, 161
    objectives, 159
    stores requisition, 164
Chester, Midlands Electricity Board
    Research Centre, 4
Cold room capacity, 20
Commercial opportunities, 28
Computer
  input documents, 161
    catering stores requisition, 164
    daily catering production return,
      163
    goods received, 162
    invoice coding, 168
    menu plan, 164
    recipe coding sheet, 165
    stock coding sheet, 166
  reports, 174
    food issues listing, 176
    food issues summary, 176
    food receipts listing, 176
    menu costing, 177
    pre-sort vetting, 174
    production unit listing, 176
    recipe cost listings, 177
    recipe costing, 176
    stock listings, 175

Computer—*contd.*
  reports—*contd.*
    transaction listing, 175
  system, 161
Conferences, 22
Consumer reaction, 23
Containers, 33
Cook–Freeze process, 32
Cooking
  conventional, 45
  times, 47
Costs
  capital, 7
  overhead, 23
  recovery, 23

Daily catering production return, 163
Darenth Park Hospital, 4
Data processing, 160

Egg dishes, 47
Enzyme action, 47

Fish dishes, 97–116
Flour and starch, 46
Food
  commercially produced, frozen, 25
  costs, 25
  critical items, 47
  distribution, 26
  issues listing, 176
  issues summary, 176
  packs, 20
  receipt listings, 176

Food—*cont.*
  service, 22
  temperatures, 20
  transit, 20
Freezer tunnel, 19
Freezing time, 19

Gelatin, 46
Goods received, 162

Hydrolysis, 46

ICI
  factory, Ardeer, Scotland, 5
  recipe manual, 14
ICL direct data entry, 161
Input documents, 161
Installation programme, 11
Insurance, 25
Invoice coding, 168

Keele University, 5
  conventional catering system, 5–6
  cook–freeze system
    assessment, 12–28
    choice of scheme, 6–7
    installation programme, 11–12
Kitchen
  capacity, 14
  techniques, 17

Leeds University, Department of
     Food and Leather
     Technology, 3
Liverpool School Meals Organisation,
  4

Meat
  changes in appearance, 47
  dishes, 51–81
  fat content, 45
Medical Officer of Health, 7

Menu
  costing, 177
  cycle, 9
  plan, 164
Micro-organisms, 47
Midlands Electricity Board Research
    Centre, Chester, 4, 7

Oxidation of fat molecules, 45

Phenolase, 47
Portioning, 47
Pre-sort vetting, 174
Production
  kitchen, 14
  levels, 27
  record, 31
  team, 15
  work hours, 17
Production Unit listing, 176

Quality and cost comparison, 23

Recipe(s)
  chicken, duck and game dishes,
    83–96
    chicken
      à la king, 85
      and ham pie, 86
      chasseur, 87
      croquettes, 88
      curried, 91
      fricassé, 89
      roast, 94
    coq au vin, 90
    duckling a l'orange, 92
    guinea fowl, roast, 95
    liver pate, 93
    pheasant, roast, 96
  cost listing, 15
  costing, 176
  farinaceous and vegetarian dishes,
    117–34
    cheese and nut pancakes, 119
    cheese and onion pasties, 120

Recipe(s)—*contd.*
farinaceous and vegetarian dishes
    —*contd.*
  cheese, potato and onion pie, 121
  egg
    croquettes, 122
    florentine, 123
  gnocchi
    gratinée, 124
    in tomato sauce, 125
  macaroni au gratin, 126
  panhaggerty, 127
  pizza, 128
  ratatouille, 129
  risotto, 130
  spaghetti au gratin, 131
  stuffed aubergines, 132
  stuffed marrow, 133
  sweetcorn croquettes, 134
fish dishes, 97–116
  American fish pie, 99
  crawfish, American style, 100
  fillet of fish
    bonne femme, 101
    bretonne, 102
    duglère, 103
    in shrimp sauce, 105
    portugaise, 104
  fish
    florentine, 106
    fried in batter, 111
    fried in breadcrumbs, 112
    in parsley sauce, 107
    mornay, 108
    veronique, 109
    waleska style, 110
  goujons of sole, 113
  salmon mousse, 114
  scampi American style, 115
  sole dieppoise, 116
ingredients, 45
meat dishes, 51–81
  beef
    carbonnade of, 58
    curried, 61
    hot pot, 53
    roast, 68
    stroganoff, 54

Recipe(s)—*contd.*
meat dishes—*contd.*
    braised liver and onions, 55
    braised steak, 56
    breaded lamb chop, 57
    cassoulet, 59
    cottage pie, 60
    Irish stew, 62
    liver and bacon pie, 63
    meat, potato and onion pie, 64
    minced beef and onions, 65
    moussaka, 66
    noisette of lamb chasseur, 67
  roasts
    beef, 68
    leg of lamb, 69
    loin of pork, 70
    saddle of lamb, 71
    sirloin of beef, 72
  sausage lyonnaise, 73
  sauté'd beef and carrots, 74
  sauté'd beef and mushrooms, 75
  steak and kidney
    pie, 76
    pudding, 77
  steak and mushroom pie, 78
  steak and onion pie, 79
  stuffed shoulder of lamb, 80
  stuffing for pork, 81
method, 17
modification, 48
standardisation, 48
sweets, 135–55
  apple and blackberry tart, 138
  apple and blackcurrant pie, 139
  apple tart, 137
  apricot
    crumble, 140
    tart, 141
  baked plum sponge, 142
  bakewell tart, 143
  bread and butter pudding, 144
  crumble mixture, 145
  Dutch apple tart, 146
  pear upside down, 147
  pineapple upside down, 148
  rhubarb and ginger tart, 149
  rice pudding, 150

Recipe(s)—*contd.*
　sweets—*contd.*
　　sago pudding, 151
　　steamed marmalade sponge, 152
　　steamed syrup sponge, 153
　　sweet shortcrust pastry, 155
　　syrup tart, 154
　testing, 15
Refectory outlets, 21

Site selection, 13
Specification, 8
Staff
　redundancy, 25
　semi-skilled, 45
Staffordshire Refrigeration and Air
　　　Conditioning Ltd, 13
Starch, 46
Stock coding sheet, 166
Stock listings, 175
Students
　fees, vi

Students—*contd.*
　habits, 49
　service to, v, xi

Tapioca starches, 46
Tender, 7
　evaluation, 11
Terms of reference in feasibility study,
　　3
Transaction listing, 175

Unit costs, 23

Variety of choice, 26
Vegetables
　colour change on storage, 47
　cooking, 46

Waxy maize flour, 19
Work study report, 12